AF388675

Green and Sustainable Solvents II: A Themed Issue in Honor of Professor Giovanni Sindona on the Occasion of His 70th Birthday

Green and Sustainable Solvents II: A Themed Issue in Honor of Professor Giovanni Sindona on the Occasion of His 70th Birthday

Editors

Monica Nardi
Antonio Procopio
Maria Luisa Di Gioia

MDPI • Basel • Beijing • Wuhan • Barcelona • Belgrade • Manchester • Tokyo • Cluj • Tianjin

Editors

Monica Nardi
Department of Health Sciences
Università Magna Graecia
Germaneto
Italy

Antonio Procopio
Department of Health Sciences
Università Magna Græcia
Germaneto
Italy

Maria Luisa Di Gioia
Dipartimento di Farmacia
e Scienze della Salute
e della Nutrizione
Università della Calabria
Arcavacata di Rende
Italy

Editorial Office
MDPI
St. Alban-Anlage 66
4052 Basel, Switzerland

This is a reprint of articles from the Special Issue published online in the open access journal *Molecules* (ISSN 1420-3049) (available at: www.mdpi.com/journal/molecules/special_issues/green_solvents_ II).

For citation purposes, cite each article independently as indicated on the article page online and as indicated below:

LastName, A.A.; LastName, B.B.; LastName, C.C. Article Title. *Journal Name* **Year**, *Volume Number*, Page Range.

ISBN 978-3-0365-2168-8 (Hbk)
ISBN 978-3-0365-2167-1 (PDF)

Contents

Preface to "Green and Sustainable Solvents II: A Themed Issue in Honor of Professor Giovanni Sindona on the Occasion of His 70th Birthday"

Dear Colleagues,

Prof. Giovanni Sindona has been a full professor of Organic Chemistry at the University of Calabria (Unical) since 1990, and has made seminal contributions to the areas of food chemistry, mass spectrometry, and green chemistry. He graduated in chemistry in 1972 with the highest marks and pize at the University of Messina, by discussing a thesis held at its Institute of Organic Chemistry on modern bioanalytical applications of mass spectrometry. He completed his training in 1977 at both the Universities of Messina and Calabria, followed by a stint at the Institut für Physikalische Chemie of the University of Bonn (Germany) as a fellow of the prestigious Alexander von Humboldt Foundation (of which he became a lifetime member), by perfecting the knowledge of mass spectrometry in the group of professor Hans D. Beckey. Here, he developed the knowledge of desorption methodologies, which allowed for the first time the direct characterization of proteins and nucleic acids by mass spectrometry. His formation was completed as a NATO fellow at the King's College of London, where at the beginning of 1980, under the direction of Professor Colin B. Reese, he developed new synthetic organic methods for the obtainment of nucleic acids segments with a predetermined sequence. Professor Sindona also taught organic chemistry at the new Faculty of Pharmacy of the University Magna Graecia of Catanzaro, from 1990 to 1993.

From 1990 to 2019, Prof Sindona has organized and directed five NATO International Schools for young researchers on the application of mass spectrometry to biomolecular chemistry, held in Italy and abroad. In 2009, as a member of the programme The Science for Peace and Security, he organized a NATO-ARW seminary in Calabria for experts coming from all over the world on the "Detection of Biological Agents and Toxins for the Prevention of Bioterrorism in Homeland Security by Advanced Mass Spectrometric Methods". In April 2016, as local director he organized the NATO SPS ASI 984915 event in Calabria, titled "Molecular Technologies for Detection of Chemical and Biological Agents", for young PhD students working on environmental security problems.

Professor Sindona is a coauthor of nearly four-hundred scientific publications on international papers in the English language (246) and ten book chapters edited by international societies on themes related to his scientific applications (10). He has been an invited speaker at many national and international meetings in the field of chemistry, author of national and international conferences (149) and a teacher at doctorate schools at the American University of Vanderbilt (Nashville) and Purdue (West-Lafayette) and at the Memorial University of St. John's in Canada. He is interested in the topics of food chemistry, mass spectrometry, and green synthesis and extraction methods.

In the field of green chemistry, he has obtained interesting results in the efficient and stereoselective synthesis of trans-4,5-diaminocyclopent-2-enones from furfural using green solvents, published in 2013 in *ACS Sustainable Chemistry* and in 2017 in *Green Chemistry*.

Notably, interest has been raised in the use of homogeneous and heterogeneous catalysis in eco-friendly synthesis, which led to the publication of numerous and interesting papers in major international scientific journals in the field of green chemistry and the development of new sustainable methods for the extraction of nutraceutical compounds from biomass.

Monica Nardi, Antonio Procopio, Maria Luisa Di Gioia
Editors

Article

Immobilization in Ionogel: A New Way to Improve the Activity and Stability of *Candida antarctica* Lipase B

Alfonso Escudero [1], Antonia Pérez de los Ríos [2], Carlos Godínez [1], Francisca Tomás [2] and Francisco José Hernández-Fernández [2,*]

[1] Department of Chemical and Environmental Engineering, Technical University of Cartagena (UPCT), Campus La Muralla, C/Doctor Fleming S/N, E-30202 Cartagena, Murcia, Spain; alfonso.escudero@upct.es (A.E.); carlos.godinez@upct.es (C.G.)

[2] Department of Chemical Engineering, Faculty of Chemistry, University of Murcia (UMU), P.O. Box 4021, Campus de Espinardo, E-30100 Murcia, Spain; aprios@um.es (A.P.d.l.R.); ptomas@um.es (F.T.)

* Correspondence: fjhernan@um.es; Tel.: +34-868889758

Academic Editor: Monica Nardi

Received: 15 June 2020; Accepted: 10 July 2020; Published: 15 July 2020

Abstract: New *Candida antarctica* lipase B derivatives with higher activity than the free enzyme were obtained by occlusion in an organogel of an ionic liquid (ionogel) based on the ionic liquid [Omim][PF$_6$] and polyvinyl chloride. The inclusion of glutaraldehyde as a crosslinker improved the properties of the ionogel, allowing the enzymatic derivative to reach 5-fold higher activity than the free enzyme and also allowing it to be reused at 70 °C. The new methodology allows enzymatic derivatives to be designed by changing the ionic liquid, thus providing a suitable microenvironment for the enzyme. The ionic liquid may act on substrates to increase their local concentration, while reducing water activity in the enzyme's microenvironment. All this allows the activity and selectivity of the enzyme to be improved and greener processes to be developed. The chemical composition and morphology of the ionogel were also studied by scanning electron microscopy–energy dispersive X-ray spectroscopy, finding that porosity, which was related with the chemical composition, was a key factor for the enzyme activity.

Keywords: enzymatic immobilization; organogels; ionogels; ionic liquid; ester synthesis; enzyme; green chemistry

---

1. Introduction

In recent decades, ionic liquids have demonstrated their potential for use as reaction and separation media. For example, in separation applications, they have been used as liquid–liquid biphasic systems [1,2] and as liquid phase in supported liquid membranes and polymer inclusion membranes [3–5]. In the field of biocatalysis, ionic liquids have been used as free solvent [6–8], adsorbed [9–12], covalently linked to particle enzymes [13–15], or as polymeric ionic liquids [16] in enzyme particles, in all cases to create an adequate microenvironment for the enzyme or to improve catalytic efficiency. The advantages of using ionic liquids instead of organic solvents as reaction media for biocatalysis include their ability to enhance enzyme activity as well as their selectivity and stability [6–9,13]. From an environmental point of view, the main advantage of ionic liquids is their lack of vapor pressure, which prevents the emission of solvent vapor into the atmosphere.

For many industrial applications, enzymes must be immobilized in order to increase their stability in operational conditions and allow their reuse. Among the different methods that can be used to immobilize enzymes, encapsulation is of particular interest because of its simplicity. The encapsulation process is based on entrapping the enzyme in a polymer matrix (gel), and there is no covalent association between the network and the enzyme [17]. Other advantages include the permeability of the matrices, which can be tuned to increase the selectivity of the biocatalyst. Very recently, de los Rios´ group have developed a new method for the enzymatic immobilization of laccase by entrapment in ionogel (a gel based on ionic liquids). Initially, (1-octyl-3-metylimidazolium bis(trifluoromethylsulfonyl)imide, cholinium bis(trifluoromethylsulfonyl)imide, cholinium dihydrogenphosphate, and hydroxyethylammonium formate were studied as choices for the constituents of the active phase of the ionogel. Using the new formulation, the enzyme's activity and stability were dramatically improved, and when the new enzymatic derivatives were applied in a batch reactor to decolor the anthraquinonic dye Remazol Brilliant Blue R until 80% decoloration was obtained [18,19].

In this work, a new immobilization method was applied to *Candida antarctica* lipase B (CaLB) to synthesize butyl butyrate as a model reaction for the synthesis of esters. For this purpose, PVC (polyvinyl chloride) was used as polymer in which a suspension of the ionic liquids with the enzyme was immobilized by entrapment. As ionic liquid, 1-octyl-3-metylimidazolium hexafluorophosphate was used because it has been demonstrated to be a suitable reaction medium for the biocatalyst CaLB [20]. The activity at different temperatures and operational stability at high temperature of this new enzymatic ionogel were analyzed.

2. Materials and Methods

2.1. Enzyme and Chemicals

A commercial lipase (EC 3.1.1.3) preparation was used as catalyst: free CaLB aqueous solution (lipozyme CaLB L), which was a gift from Novo España S.A. Polyvinyl chloride powder (PVC) was purchased from Sigma-Aldrich. 1-Octyl-3-methylimidazolium hexafluorophosphate ([Omim][PF$_6$]) was purchased from IOLITEC (purity > 99%).

2.2. Immobilization of Free CaLB in Gels Based on Ionic Liquids (Ionogel)

Free CaLB was entrapped by occlusion in an ionogel based on [Omim][PF$_6$]. Derivatives of ionogel enzymes were prepared using 50% PVC and 50% IL using the following general procedure: 1 mL of THF was added to 200 µL of ionic liquid [Omim][PF$_6$] and the mixture was stirred. Subsequently, 100 µL of a 14.2 mg/mL concentration CaLB solution or that four-fold diluted (3.55 mg/mL) in phosphate buffer (20 mM, pH = 7) was added to the IL mixture dissolved in THF under shaking conditions. Then, after adding 100, 20, or 0 µL of glutaraldehyde (25% water solution), 200 mg of PVC were added very slowly and continuously shaken. One additional milliliter of THF was added to allow the PVC solution. The process ended when the solution was seen to be homogeneous. The magnetic stirrer was removed, and the previous solution was poured into the center of an O-ring which was placed on a glass plate. The mixture was left to stand for 48 hours in a suction flow chamber to vaporize the THF while the PVC occluded the ionic liquid and enzyme it contained. After drying, the mixture was seen to solidify on the glass ring plate. The ring was then removed to release the circle of ionogel CaLB derivative, which was crushed for use in the reactors (Figure 1). The different enzymatic derivatives prepared are summarized in Table 1.

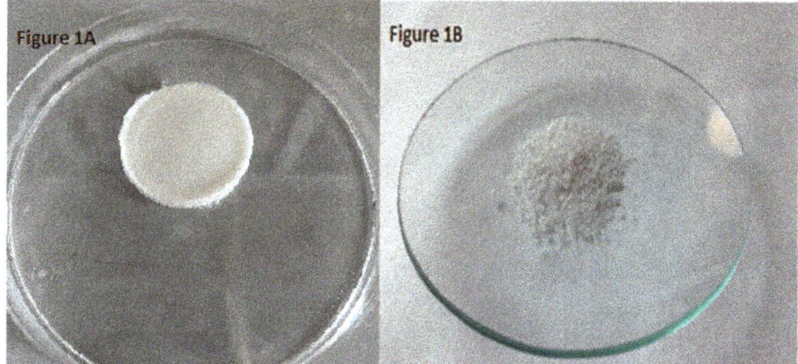

Figure 1. (**A**) Ionogel enzymatic derivative after casting, (**B**) Enzymatic derivative after crushing.

Table 1. Chemical composition of the prepared ionogel enzymatic derivatives. The "ionogel" in the name of the immobilized enzyme stands for PVC + IL.

	Enzyme (µL)	Ionic Liquid [Omim][PF$_6$] (mg)	PVC (mg)	Glutaraldehyde 25% in Water (µL)	[Protein in the Reactor], 80 mg of Ionogel Was Used
Enz [a]	100 (14.2 mg prot/mL)	0	0	0	1.420
Enz-PVC [b]	100 (14.2 mg prot/mL)	0	200	0	0.379
Enz-PVC-B [c]	100 (3.55 mg prot/mL) *	0	200	0	0.095
Enz-ionogel [d]	100 (14.2 mg prot/mL)	200	200	0	0.227
Enz-ionogel-B [e]	100 (3.55 mg prot/mL) *	200	200	0	0.057
Enz-ionogel-20G [f]	100 (14.2 mg prot/mL)	200	200	20	0.218
Enz-ionogel-100G [g]	100 (14.2 mg prot/mL)	200	200	100	0.189
Enz-ionogel-B-20G [h]	100 (3.55 mg prot/mL) *	200	200	20	0.055
Enz-ionogel-B-100G [i]	100 (3.55 mg prot/mL) *	200	200	100	0.047

* Initial enzyme concentration diluted four-fold with phosphate buffer., [a] Free Enzyme., [b] PVC enzymatic derivative without IL., [c] PVC enzymatic derivative with buffer (B) and without IL., [d] Ionogel enzymatic derivative without buffer [e] Free Enzyme., [e] Ionogel enzymatic derivative and buffer., [f] Ionogel enzymatic derivative and 20 µL glutaraldehyde and without buffer., [g] Ionogel enzymatic derivative and 100 µL glutaraldehyde and without buffer., [h] Ionogel enzymatic derivative and 20 µL glutaraldehyde and with buffer., [ij] Ionogel enzymatic derivative and 100 µL glutaraldehyde and with buffer (B).

2.3. Ionogel-Derived Enzymatic Activity, Conversion, and Stability

Butyl butyrate synthesis from vinyl butyrate and butanol was used as reaction model (Figure 2) for the determination of activity, stability, and final conversion. For this, 0.5 mL (300 mmol) vinyl butyrate in hexane and 0.5 mL (300 mmol) 1-butanol in hexane were added to 2 mL screw-capped vials. The reaction was started by adding 80 mg ionogel CaLB derivative or free enzyme (1.42 or 0.355 mg protein) and allowed to run, whilst stirring, for 300 min at 30 °C. At different times, 30 µL aliquots were suspended in 920 µL hexane with 50 µL of valeric acid (100 mM, internal standard). The resulting solution (5 µL) was analyzed by GC. All experiments were carried out in duplicate, and the mean values are reported. The standard deviation was calculated in all cases. The efficiency of the catalytic action was measured based on two parameters: (**i**) the synthetic activity, defined as the amount of mmol of butyl butyrate produced per mg of protein and per minute and (**ii**) the final vinyl butyrate conversion at 300 min. One unit (U) produced 1 µmol of butyl butyrate per minute. Enzyme stability was measured by reusing the ionogel in different cycles at 70 °C. In each cycle the enzyme derivative was suspended in hexane for 24 h. For the stability assays the activity of CaLB was measured as described above and residual activity was calculated. Free CaLB (100 µL solution; 14.2 mg/mL) was also assayed in order to compare the efficiency of free enzyme with that of the immobilized enzymes.

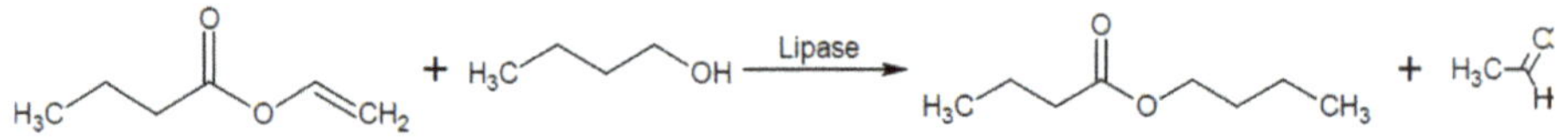

Figure 2. Synthesis of butyl butyrate from vinyl butyrate and 1-butanol catalyzed by CaLB.

2.4. Gas Chromatographic Analysis

The sample analyses were performed in a gas chromatograph (model 450 GC from Bruker, Germany) equipped with a flame ionization detector (FID), an autosampler injector, and a capillary column from J & W Scientific, Agilent Technologies, model HP INNOWAX, 30 m in length, 0.25 mm nominal diameter, and 0.25 micron film thickness. The injector temperature was 210 °C, working with a constant flow of 1 mL/min at the head of the column and split ratio of 1:10. N_2 was used as carrier gas, while the temperature in the detector was 220 °C. The temperature program was as follows: 40 °C for 1 min; increase at 15 °C min^{-1} for 1 min, increase at 35 °C min^{-1} for 3 min; hold at 160 °C for 6 min. The retention times of the peaks were as follows: vinyl butyrate, 4.0 min; 1-butanol, 4.6 min; butyl butyrate, 5 min; butyric acid, 7.9 min; valeric acid (internal standard), 9.4 min. Substrate and product concentrations were calculated from calibration curves using stock solutions of pure compounds.

2.5. SEM–EDX Analysis

Scanning electron microscopy with energy dispersive X-ray spectroscopy (SEM–EDX) was applied for microimaging of the ionogel. The SEM images were acquired using a HITACH S-3500N apparatus following the protocol used by the SAIT platform (Scientific Instrumentation Service of the Technical University of Cartagena, Spain). The same equipment was used to generate the EDX profiles.

3. Results

3.1. Activity and Conversion of Ionogel-Derivatives

Among the ionic liquids used for enzyme-catalyzed reactions, water-immiscible ionic liquids were generally found to be suitable for biocatalytic reactions, while the water-miscible ionic liquids assayed were considered worse than conventional media. In the case of most water-miscible ILs [20], the negative effect observed on the lipase activity can be attributed to the direct interaction of the anion with the enzyme molecules, which would lead to protein denaturation or stripping off the essential water associated with the enzyme [21–23]. More specifically, van Rantwijk et al. [22] observed a strong coordinating effect on the part of anions, which led to the deactivation of CaLB in [Bmim][NO$_3$], [Bmim][CH$_3$CH$_2$COO] and [bmim][dca], both of which are water-miscible ILs, and to which the dissolution of CaLB in these ILs is attributed. However, interaction between ILs and enzymes may sometimes improve enzyme behavior. The ionic liquid [Chol][H$_2$PO$_4$], which is water-soluble, was seen to be very effective at enhancing and stabilizing laccase activity, which was attributed to the modifications in the secondary structure of the enzymatic protein that it produced [24]. A modification of the secondary structure was also observed in CaLB in water-immiscible ionic liquids, which resulted in a more compact enzyme conformation that exhibited catalytic activity [25].

As mentioned above, the present work describes how CaLB was entrapped in a polymeric matrix (PVC) with the ionic liquid 1-octyl-3-metylimidazolium hexafluorophosphate ([Omim][PF$_6$]). The new enzymatic derivatives obtained were named ionogels of CaLB. In our assay, the hydrophobic ionic liquid, [Omim][PF$_6$], was used because it creates a microenvironment that is suited to CaLB, thus increasing the enzyme activity over levels that can be obtained with hexane (used here as reference organic solvent) and other imidazolium-based ionic liquids with lower alkyl chain lengths that are commonly used as reaction media in transesterification reactions [7,14,26–28]. The entrapment method involving ionic liquids allows the enzyme to be immobilized by occluding CaLB within a suitable microenvironment that allows the enzyme to be retained while allowing the substrates and products of the reaction to cross the ionic liquid layer around the enzyme. Even if the ionic liquid is a good solvent for the substrates of the reaction, the microenvironment concentration of the substrates around the catalyst could be increased, consequently increasing the kinetics of the process. All enzyme derivatives (see Table 1) were tested as catalysts for the synthesis of the butyl butyrate from vinyl butyrate and 1-butanol. As a kinetically controlled reaction catalyzed by a serine hydrolase enzyme (CaLB), the transformation of vinyl butyrate is closely dependent on the nucleophile acceptors present in the reaction medium and involves competitive distribution of the rapidly formed acyl–enzyme intermediate between water (hydrolysis) or another nucleophile reagent, such as 1-butanol (transesterification). The latter synthetic pathway can be enhanced by using activated acyl donors such as vinyl esters with very low water content in the medium and high nucleophile (e.g., 1-butanol) concentration. The efficiency of the catalytic action can be measured by two parameters, the synthetic rate and the amount of converted vinyl butyrate measured at the end of the reaction (Figure 3). Free enzyme also was tested in order to compare the results with immobilized enzyme.

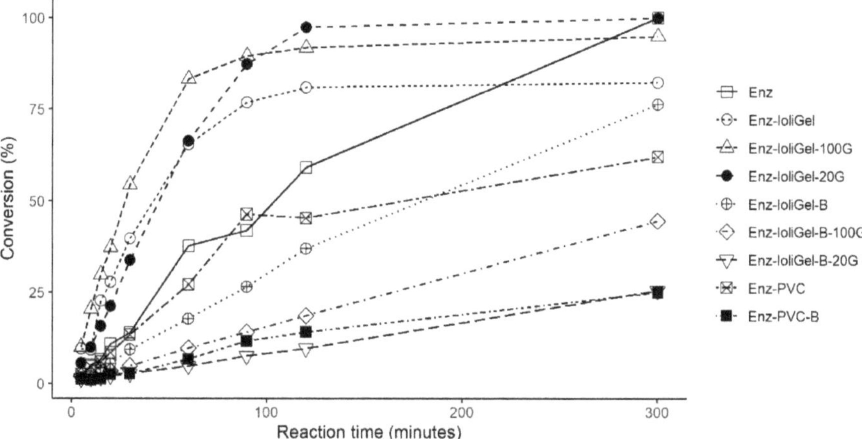

Figure 3. Time course of butyl butyrate synthesis from vinyl butyrate (150 mM) (not depicted) and 1-butanol (not depicted) (150 mM) at 30 °C. The synthesis activity was calculated from the initial slope and the conversion % was calculated at the end of the reaction (300 min).

As can be seen from Figure 4, CaLB-ionogel containing [Omim][PF$_6$] and 100 µL of glutaraldehyde had the best activity values, which are much higher than those achieved with the free enzyme. The second highest activity (half of the first one) was obtained when 20 µL glutaraldehyde was added. In the latter case (Enz-ionogel-20G), 100% conversion was reached at 300 min. Slightly lower activity (7.4 U/mg prot) was observed when neither glutaraldehyde nor buffer was added (Enz-ionogel). The activity of the free enzyme was much lower than the activity measured for the other CaLB-ionogel derivatives, except those produced with buffer and 20 µL of glutaraldehyde (Enz-ionogel-B-20G). It should be recalled here that "ionogel" in the reference name of the immobilized enzyme form stands for PVC+IL. The lowest activity value was that obtained when the diluted enzyme (with buffer) was immobilized in PVC without ionic liquid. By comparing the different activity and conversion values of the enzyme derivatives, it can be seen that the best activity and conversion values were obtained for the CaLB-ionogel derivative when enzymes undiluted (i.e., without buffer) were used: compare Enz-ionogel-B-100G vs. Enz-ionogel-100G, Enz-ionogel-B-20G vs. Enz-ionogel-20G, Enz-PVC-B vs. Enz-PVC). Hence, the higher the enzyme concentration in the enzymatic derivative, the higher the activity of the enzyme per mg of protein. Furthermore, increasing the amount of glutaraldehyde results in increasing enzyme activity: compare Enz-ionogel-100G vs. Enz-ionogel-20G and Enz-ionogel-B-100G vs. Enz-ionogel-B-20G, in which increasing the amount of glutaraldehyde used from 20 to 100 µL doubles the enzyme activity. This fact highlights the important role of glutaraldehyde in the enzymatic derivative. Sheldon developed a new method for enzyme immobilization as CLEA (crosslinked enzyme aggregates). The method involves the precipitation of the enzyme from aqueous buffer followed by crosslinking with glutaraldehyde of the resulting physical aggregates of enzyme molecules [29,30]. CLEAs are shown to be stable, recyclable catalysts exhibiting high catalyst productivities. For instance, *C. antarctica* lipase B, adsorbed and crosslinked on a polypropylene carrier, maintained its transesterification activity in the ionic liquids [BMIm][NO$_3$] and [BMIm][dca], which deactivate the free enzyme [6]. Perhaps the formation of enzyme aggregates in ionogel derivatives by crosslinking with glutaraldehyde involves improvement of the enzymatic derivatives' properties.

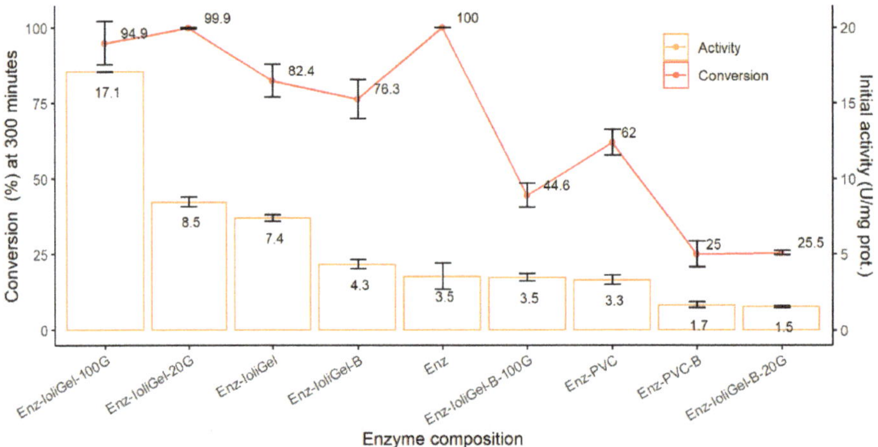

Figure 4. Activity and conversion exhibited by CaLB for butyl butyrate synthesis for different immobilized CaLB and free enzyme at 30 °C.

Note that the Enz-ionogel-100G reaches 5-fold higher activity than the free enzyme. It is also important to point out that in the case of Enz-PVC derivative, the reaction nearly stops at 40% conversion. It could be explained by possible PVC degradation at these conditions and, consequently, releasing HCl into the medium. Very recently, lipase degradation of plasticized PVC has been studied [29,30]. It was reported that this degradation could yield HCl. All the enzymatic derivatives, except Enz-PVC, are made of ionic liquids or contain buffer. The buffer could reduce the HCl concentration and the ionic liquids could reduce the effective HCl concentration. Therefore, the HCl could affect the enzyme activity only in the case of Enz-PVC. Furthermore, the ionogel could reduce PVC degradation.

In our previous work, laccase was immobilized in an ionogel based on 50% PVC and 50% the ionic liquid 1-octyl, 3-methyl imidazolium bistrifluoromethyl sulfonyl imide [Omim][NTf$_2$]. To increase the activity and stability of the enzymatic derivative, glutaraldehyde was added at different concentrations (between 0.25% and 1.5%). An increase in activity and stability was observed at all the assayed concentrations, with the maximum being 0.5% glutaraldehyde [19]. In our experiments with CaLB, an increase in activity was observed at all assayed glutaraldehyde concentrations (between 1% and 4%) at 30 °C when the enzyme was used without dilution by buffer. The use of enzyme diluted with buffer decreased the activity with respect to non-diluted medium, probably because the diluted enzyme hindered crosslinking with glutaraldehyde. In the present work, the highest lipase activity was reached at 4% glutaraldehyde concentration without enzyme dilution, as mentioned. The effect observed for glutaraldehyde in both cases (laccase and lipase) shows the importance of the crosslinking to enhance the activity of the immobilized derivative.

We can also observe that the immobilization of the enzyme on a PVC support without ionic liquid provided the lowest activity values. Furthermore, the comparison of enzymatic derivatives with and without ionic liquids (i.e., Enz-PVC-B vs. Enz-ionogel-B and Enz-PVC vs. Enz-ionogel) demonstrates the importance of using ionic liquid in the preparation of the immobilized derivatives. As mentioned above, the increase in activity could be due to the suitable microenvironment created by the ionic liquid around the enzyme and/or to increased substrate concentration in the ionic liquid microenvironment around the enzyme. In this immobilization method, the enzyme is entrapped by the ionic liquid and PVC, and the ionic liquid could absorb substrates, increasing the substrate concentration in the microenvironment of the enzyme. In this respect, the distribution ratio of vinyl butyrate, butanol, and butyl butyrate between ionic liquid and hexane was studied by Hernández-Fernández et al. 2010 [2]. The distribution ratio of vinyl butyrate, butanol, and butyl butyrate between [Omim][PF$_6$] and hexane was 0.68, 3.63, and 0.35, respectively. These distribution ratios could increase the concentration of butanol in the microenvironment of the enzyme and remove the product butyl butyrate from this microenvironment. All the above might improve the kinetics of the bioreaction. Moreover, it has been demonstrated [12] that enzyme immobilization involves more stable enzymes in exchange of less enzymatic activity. As mentioned above, the major advantage of this method is that it allows more active derivatives to be created than in the case of free enzyme, as can be seen observed from Figure 4.

Regarding the selectivity of the reaction, the byproduct butyric acid was not observed in the enzymatic derivatives obtained using ionic liquids. The synthetic pathway was enhanced by the use of activated acyl donors (vinyl butyrate), a medium with a very low water content (hexane), and a strong nucleophile (1-butanol). Furthermore, the ionic liquid has a specific ability to reduce water activity in the enzyme microenvironment by capturing the free molecular water and, consequently, reducing the hydrolysis reaction. The increased selectivity of transesterification reactions using ionic liquids has been previously described [9,20].

As regards the greenness of the process, we should remember that the primary cause of the high E factors (developed by Seldon and defined as the ratio mass waste/mass product) of most processes in the pharmaceutical industry is the high molecular complexity of APIs (active pharmaceutical ingredients), the large number of chemical steps needed to assemble the APIs from commercially available starting materials, and the use of classical stoichiometric reagents instead of catalysts [29,31]. From an environmental point of view, the new method proposed in this paper involves the use of immobilized biocatalysts with better properties, activity, and selectivity than the free enzyme, which increases the productivity of the process [30,32]. Furthermore, the ionic liquid that traps the enzyme not only improves the catalytic activity of the biocatalyst but also concentrates the substrate around the catalyst, increasing the kinetics of the reaction. The ionic liquid around the catalyst was also able to acts as a selective membrane between the medium and the biocatalyst, regulating the transport of substrates towards the catalyst and, consequently, improving the overall selectivity of the process. The consequence of all this is that number of the purification steps could be reduced, thereby making the whole process greener.

3.2. Stability of Ionogel Enzymatic Derivatives

The stability of the new enzymatic derivatives was measured by reusing the catalyst in 5-day cycles. Figure 5 shows the activity of the first cycle using the ionogel of CaLB and the free enzyme at 70 °C. The relative activity between the different immobilized derivatives was similar at 30 °C, which corroborates the results found for the new enzymatic derivatives. Furthermore, in all cases, the activity increased two- to three-fold by increasing temperature because the kinetic constant increased with temperature. However, the conversion was higher in some cases and lower in others—while higher initial activity could lead to a higher final conversion %, a higher temperature might also involve lower final conversion % due to deactivation of the enzyme. It is important to point out that in nearly all cases, the activity of the undiluted immobilized enzyme was higher than that of the free enzyme. Furthermore, at 70 °C, it was observed that using the concentrated enzyme in the immobilized derivative led to better activity and final conversion values than when the enzyme diluted in phosphate buffer was used. The same behavior was observed with increasing concentrations of glutaraldehyde, since activity was higher with 100 µL than with 20 µL of glutaraldehyde. As mentioned above, the same effect of glutaraldehyde concentration on the stability of immobilized laccase was observed in a previous work, when maximum stability was found at a 0.5% glutaraldehyde concentration [19].

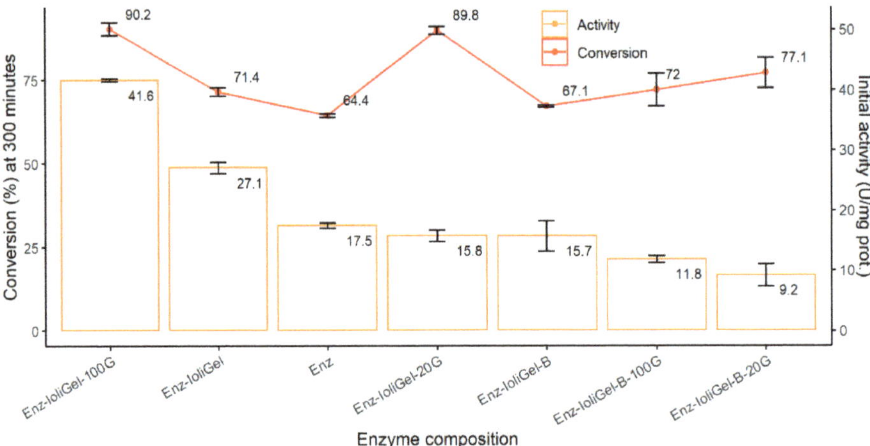

Figure 5. Activity and conversion exhibited by CaLB during butyl butyrate synthesis using different immobilized CaLB (ionogel) and free enzyme at 70 °C.

To analyze the operational stability of the new enzyme derivatives at a high temperature (70 °C), six cycles of 24 h each were carried out using the same enzyme. Only the ionogel composed of concentrated enzyme (without buffer) and 100 or 20 µL of glutaraldehyde remained active after the first cycle at 70 °C, while the rest of the enzyme derivatives were inactivated after the first assay. Figure 6 depicts the stability of Enz-ionogel-20,100G vs. temperature, represented as a decay of the initial activity. The half -life of the ionogel with 100 µL glutaraldehyde and with 20 µL glutaraldehyde was similar for around 3 cycles, which means for 60–72 h. The half-life was calculated for these two derivatives considering that each cycle lasted 24 h.

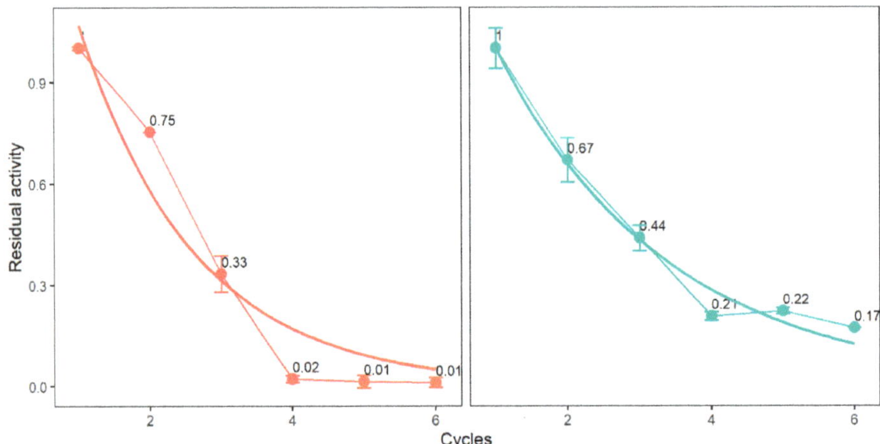

Figure 6. Deactivation profile of ENZ-ionogel-20,100G in hexane at 70 °C.

As example, Jiang et al. (2009) immobilized *Candida rugosa* lipase on magnetic nanoparticles supported ionic liquids with different cation chain lengths and anions, Cl⁻, BF_4^-, and PF_6^- [10].

The activity of bound *Candida rugosa* lipase was higher (118.3%) than that of the native lipase. The activity of *Candida rugosa* lipase immobilized on magnetic nanoparticle-supported ionic liquids reached a maximum at 37 °C, after which it decreased, remaining at a constant 60% when the temperature reached 80 °C. For ionogel-immobilized CaLB, as mentioned above, the activity at 70 °C increased two- to three-fold over the activity observed at 30 °C, which was explained by the increasing kinetic constant with temperature. *Candida rugosa* lipase retained 60% of its initial activity after eight batch reactions. Considering that each cycle lasted 5 hours, the lipase retained 60% of its activity at 40 h. As mentioned above, similar results were obtained with some ionogel CaLB derivatives, with the half-life of the ionogel with 100 and 20 μL glutaraldehyde being around 60–72 h. Another example related to CaLB analyzed the enzyme immobilized by adsorption onto 12 different modified silica supports with different alkyl chain lengths and functional groups. Immobilized enzyme particles were coated with ionic liquids (butyltrimethylammonium bistriflimide or trioctylmethylammoniun bistriflimide) by adsorption. The immobilized derivatives were assayed for the kinetic resolution of *rac*-1-phenylethanol in both ionic liquid/hexane and ionic liquid/supercritical carbon dioxide biphasic media. The best results were obtained for the supports modified with non-functionalized alkyl chains. The activity of the immobilized enzyme particles coated with ionic liquids was lower in hexane than when using the immobilized enzyme with no such coating, which was explained by the stronger mass transfer limitation caused by the ionic liquids. However, half-life times were enhanced in hexane media at 95 °C [31,33]. Porcine pancreatic lipase (PPL) was also immobilized using ionic liquids for that magnetic chitosan nanocomposites modified with a functional imidazolium-based IL. The activity of almost every PPL derivative was higher than that of the free enzyme when working above 50 °C. The thermal stability of the PPL derivatives at 50 °C ranged from 45% to 80% at 6 h [14].

3.3. SEM–EDX Characterization of the Ionogel Enzymatic Derivative

The ionogel enzymatic derivatives, which are made of ionic liquid 1-octyl-3-methylimidazolium hexafluorophosphate, PVC, and enzyme, were analyzed by SEM–EDX, after and before being crushed, and after and before being used.

Figure 7 depicts SEM micrographs of samples with buffer-diluted enzyme (without crushing) (Enz-ionogel-B), fresh and after 1 and 5 reaction cycles (24 h each) at 30 °C. Notable morphological differences can be observed between the fresh ionogel enzymatic derivative (Figure 7) and the same ionogel enzymatic derivative after 1 and 5 reaction cycles (Figure 7b,c, respectively). The irregularly distributed grains of Figure 7a change completely to show well-defined grains at the end of the first cycle (Figure 7b), a morphology that is maintained at the end of the 5 cycles (Figure 7c). This change in morphology might be due to the interaction of the PVC in the ionogel with the hexane reaction medium, giving rise to the "bubbles" that practically cover the surface of the ionogel as soon as it is used.

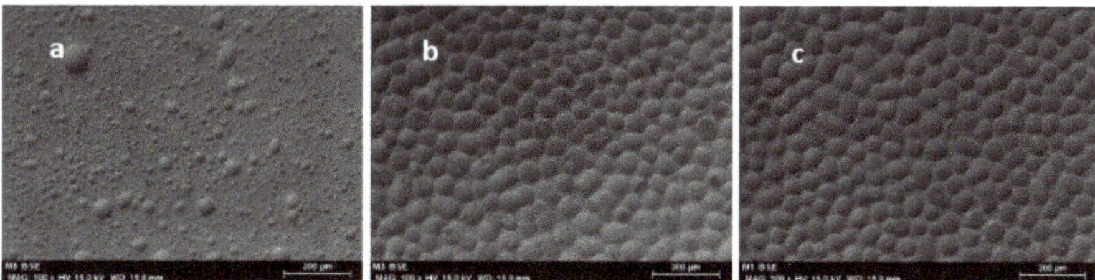

Figure 7. Scanning electron micrographs of the ionogel enzymatic derivative without crushing: (Enz-ionogel-B) (100×): (**a**) fresh (scale bar = 300 μm), (**b**) after 1 cycle (scale bar = 300 μm), and (**c**) after 5 cycles (scale bar = 300 μm).

As regards the composition of the fresh ionogel enzymatic derivative (without crushing), the EDX spectrum (Figure 8) shows that it has maintained its stability during all the operation cycles, as is evident from the superimposed peaks of the representative elements (peak F represents [Omim][PF$_6$] and peak Cl represents PVC). It should be noted that the concentration of fluorine measured in the selected samples increases with the number of reaction cycles, from 6.59% (fresh) to 7.94% (1 cycle) and, finally, 11.51% (5 cycles). This increase suggests that the IL in the ionogel migrates outward from it during the course of the reaction. It should be noted that EDX provides information on the composition of elements at a sampling depth of 1–2 μm.

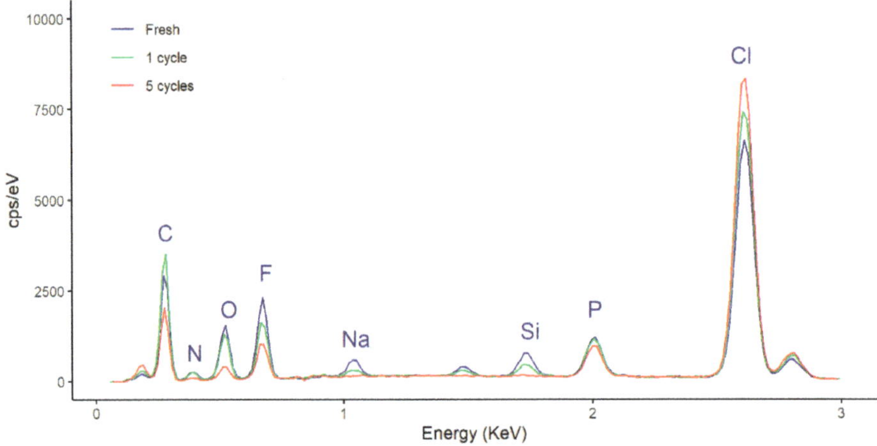

Figure 8. EDX spectra of the ionogel enzymatic derivative (Enz-ionogel-B): fresh (blue line), after 1 cycle (green line), and after 5 cycles (red line).

The SEM–EDX spectra of ionogel prepared with glutaraldehyde (Enz-ionogel-B-20,100G) were also analyzed. The glutaraldehyde was added to the ionogel in order to achieve crosslinking of the enzyme units by allowing the functional groups of the glutaraldehyde to react with the free amino groups of the enzymes, thus forming Schiff bases. This crosslinking provides stability to the enzyme derivative, as mentioned above.

Figure 9 shows the EDX spectra corresponding to fresh ionogel with diluted enzyme incorporating glutaraldehyde in the form of a 25% aqueous solution (20 or 100 μL). It can be seen that the carbon peak was significantly higher in the ionogel with 100 μL of glutaraldehyde than when 20 μL was used. Comparing the SEM micrographs of both ionogel enzymatic derivatives (Figure 10a,b), it can be seen than the ionogel enzymatic derivative made with 100 μL is slightly more porous than with 20 μL. This circumstance favors an increase in catalytic activity, due to the greater active contact surface between the enzyme and substrates. Thus, the higher the concentration of glutaraldehyde, the greater the crosslinking and porosity and, consequently, the greater the specific activity and conversion achieved. This is corroborated by comparing the activities of the respective ionogels. In the first case, the specific activity reached was 3.5, and in the second, 1.5, with respective conversions of 45% and 26%. In a previous work with laccase, the group led by de los Rios proposed that the activity and stability of the enzyme are related to the carrier's porosity and the hydrophobicity of the ionic liquids [8]. Also, a very recent work demonstrates that the addition of glutaraldehyde to the mixture of laccase–PVC–ionic liquid could contribute not only to the crosslinking of the enzyme but also to modification of the macrostructure of the membrane surfaces [19].

However, when ionogel with buffer and without glutaraldehyde was used, the activity was higher than in the case of ionogel with buffer and glutaraldehyde (20 or 100 μL). It can be hypothesized, in this case, that the addition of glutaraldehyde in the presence of a buffer provides an "extra" degree of dilution to the enzyme, which would harm the enzymatic activity. It is important to point out that the ionogel was crushed before use. When ionogels are crushed (Figure 10a,b) their appearance changes (Figure 7a, without crushing). Before use, the ionogels were crushed with a grinder, which might cause an increase in temperature and major surface restructuring due to this thermal stress.

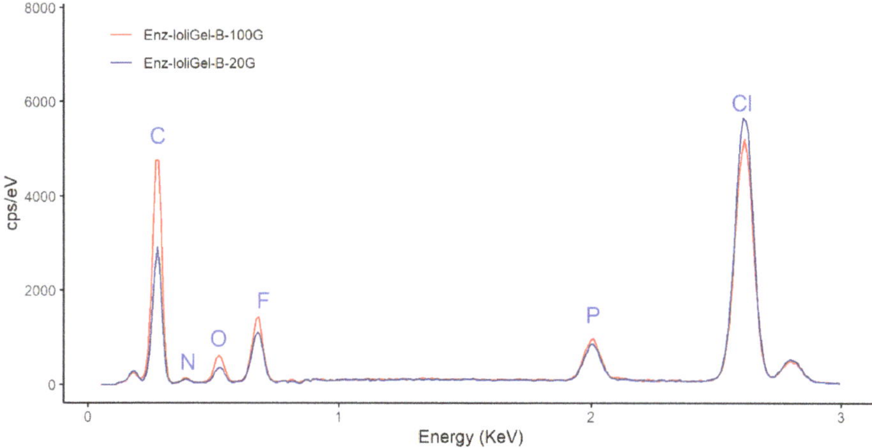

Figure 9. EDX spectra of the ionogel enzymatic derivative (Enz-ionogel-B-20,100G): 100 μL glutaraldehyde (red line) and 20 μL glutaraldehyde (blue line).

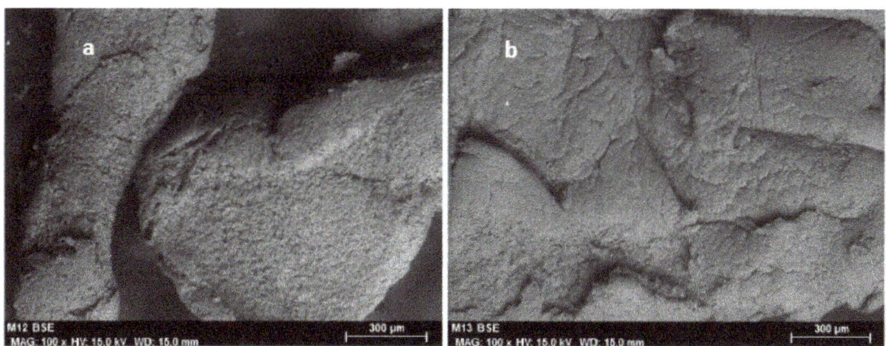

Figure 10. Scanning electron micrographs of the ionogel with diluted enzyme incorporating glutaraldehyde (25% aqueous solution) enzymatic derivative (Enz-ionogel-B-20,100G): (**a**) 100 μL glutaraldehyde (100×) (scale bar = 300 μm) and (**b**) 20 μL glutaraldehyde (100×) (scale bar = 300 μm).

Samples prepared only with PVC and ionic liquid and without enzyme were analyzed both fresh and after 5 cycles of operation. Figure 11 compares the EDX spectra of the whole and crushed fresh samples. There is practically complete overlapping of peaks, except for chlorine, which is more abundant in the whole sample, perhaps because of accumulation on some surfaces. The most relevant aspect is the total homogeneity that can be observed in the SEM micrographs (Figure 11), which do not show any relief in the intact (Figure 11 blue line) or crushed (Figure 11 green line) samples. This completely smooth surface was only observed in the absence of enzyme, when its presence makes the membrane appreciably more porous, as can be seen in Figure 7a. This may be due to water being introduced with the enzyme solution.

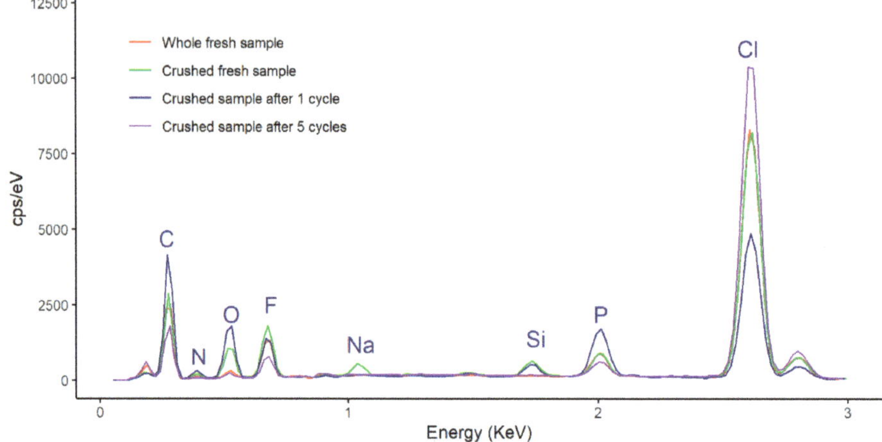

Figure 11. EDX spectra of the sample prepared with PVC and ionic liquids [Omim] [PF$_6$] in the absence of enzyme: whole fresh sample (red line), crushed fresh sample (green line), crushed sample after 1 cycle (blue line), and crushed sample after 5 cycles (purple line).

Figure 11; Figure 12 depict study the stability of the sample made of PVC and [Omim] [PF$_6$] as measured by SEM–EDX. Note that the proportion of ionic liquid (fluorine peak) decreases significantly from the fresh sample (Figure 11, 25.41%) to the samples used for 1 cycle (Figure 11, 7.73%) and 5 cycles (Figure 11, 9,16%). The loss of IL when enzyme is not used in the formulation clearly contrasts with the increased peak observed for the ionogel with enzyme (Enz-ionogel-B, see Figure 8). The explanation lies in the increase in hydrophilicity of the IL (which is normally hydrophobic) when the enzyme derivative is incorporated dissolved in water, thus reducing the possibility of migration to the hexane (hydrophobic) reaction medium. The corresponding SEM micrographs (Figure 12) show the superficial change that takes place in the samples, such as the fresh sample (Figure 12a,b). As can been observed, the smooth surface (Figure 12a,b) changes completely to show well-defined grains (Figure 12c,d). A similar phenomenon occurred with the ionogel enzymatic derivatives (see Figure 7), which was due to hexane causing the PVC to swell.

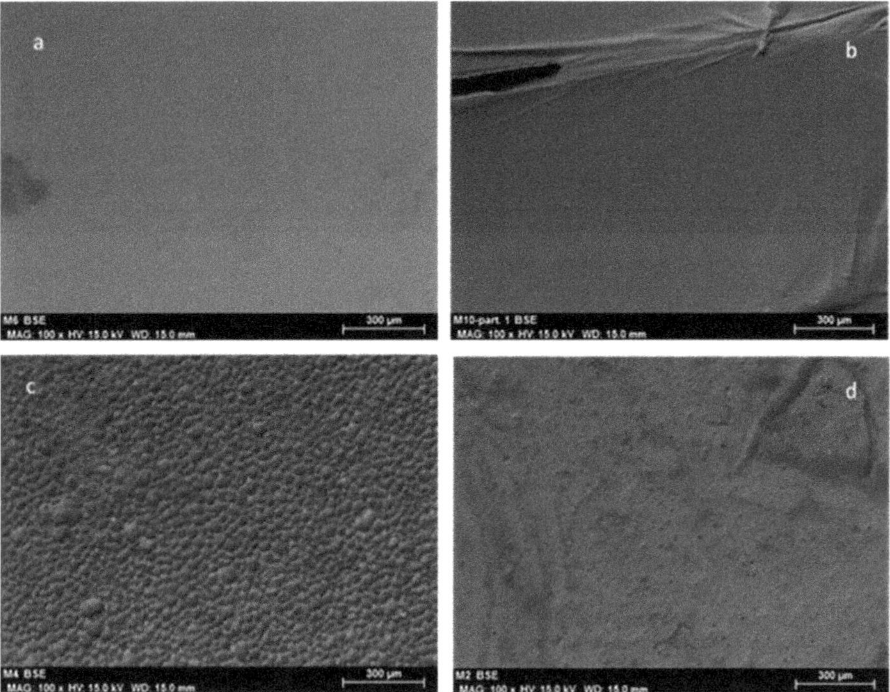

Figure 12. Scanning electron micrographs of the sample prepared with PVC and ionic liquids [Omim] [NTf$_2$] in the absence of enzyme: (**a**) whole fresh sample (100×) (scale bar = 300 μm), (**b**) crushed fresh sample (100×) (scale bar = 300 μm), (**c**) crushed sample after 1 cycle (100×) (scale bar = 300 μm), and (**d**) crushed sample after 5 cycles (100×) (scale bar = 300 μm).

4. Conclusions

A new method for immobilizing enzymes by entrapment in a gel based on an ionic liquid (ionogel) is described for CaLB. The use of the CaLB-ionogel based on the ionic liquid [Omim] [PF$_6$] and the organic polymer PVC led to higher enzymatic activities than when free CaLB was used. The use of glutaraldehyde as crosslinker agent improved the activity of the enzymatic derivative, with the highest activity being reached with PVC, [Omim][PF$_6$] ionic liquid, CaLB, and 100 μL of glutaraldehyde. The operational stability of all the enzymatic derivatives (named CaLB-ionogel) was tested at high temperatures (70 °C). Only the CaLB-ionogels prepared with glutaraldehyde were still active after 2 cycles (24 hour each cycle). The porosity of the catalyst was found to be a key factor for the activity of the enzyme. Furthermore, the ionic liquid used could play a double role: on the one hand, creating a suitable microenvironment for the enzyme, thereby increasing its activity and stability and, on the other hand, increasing the substrate concentration in the enzyme microenvironment by concentrating in the ionic liquids, thus avoiding limitations in mass transfer and increasing the kinetics of the reaction. If the ionic liquids are correctly chosen, the ionic liquids could also act as a selective liquid membrane around the biocatalyst for different substrates, improving the overall selectivity of the process. All mentioned properties demonstrate that the new biocatalyst makes processes greener. The results of this study are quite encouraging and suggest a new way to easily design a biocatalyst. By changing the ionic liquid that was used, the activity, and even the selectivity of the catalyst could be modified. For instance, to increase the activity and selectivity of a given biocatalyst for a specific substrate, it is simply a matter of choosing an ionic liquid that makes a suitable microenvironment for the enzyme and, also, in which the specific substrate will be preferentially absorbed. The method increases the potential field of application for existing biocatalysts.

Author Contributions: Conceptualization, F.J.H.-F., A.P.d.l.R. and A.E.; methodology, F.J.H.-F. and A.E.; software, C.G. and A.E.; validation, A.E., F.T. and C.G.; formal analysis, F.J.H.-F.; investigation, A.E.; resources, A.P.d.l.R.; data curation, A.E.; writing—original draft preparation, F.J.H.-F. and A.E.; writing—review and editing, F.J.H.-F. and A.P.d.l.R.; visualization, F.T.; supervision and A.P.d.l.R.; project administration, F.J.H.-F. and A.P.d.l.R.; funding acquisition, F.J.H.-F. and A.P.d.l.R. All authors have read and agreed to the published version of the manuscript.

Funding: This research was partially supported by the Spanish Ministry of Economy and Competitiveness (MINECO) (Grant number: RTI2018-099011-B-I00) and Séneca Foundation (Grant number: 20957/PI/18).

Conflicts of Interest: The authors declare no conflict of interest.

References

1. de los Ríos, A.P.; Hernandez-Fernandez, F.J.; Lozano, L.J.; Sanchez, S.; Moreno, J.I.; Godinez, C. Removal of metal ions from aqueous solutions by extraction with ionic liquids. *J. Chem. Eng. Data* **2009**, *55*, 605–608. [CrossRef]

2. Hernández-Fernández, F.J.; de los Ríos, A.P.; Gómez, D.; Rubio, M.; Víllora, G. Selective extraction of organic compounds from transesterification reaction mixtures by using ionic liquids. *AIChE J.* **2010**, *56*, 1213–1217.

3. de los Ríos, A.P.; Hernández-Fernández, F.J.; Lozano, L.J.; Sánchez-Segado, S.; Ginestá-Anzola, A.; Godínez, C.; Tomás-Alonso, F.; Quesada-Medina, J. Selective extraction of organic compounds from transesterification reaction mixtures by using ionic liquids. *J. Membrane Sci.* **2013**, *444*, 469–481.

4. Hernández-Fernández, F.J.; de los Ríos, A.P.; Tomás-Alonso, F.; Gómez, D.; Víllora, G. On the development of an integrated membrane process with ionic liquids for the kinetic resolution of rac-2-pentanol. *J. Membrane Sci.* **2008**, *314*, 238–246.

5. Baicha, Z.; Salar-García, M.J.; Ortiz-Martínez, V.M.; Hernández-Fernández, F.J.; de los Ríos, A.P.; Maqueda-Marín, D.P.; Collado, J.A.; Tomás-Alonso, F.; El Mahi, M. On the selective transport of nutrients through polymer inclusion membranes based on ionic liquids. *Processes* **2019**, *7*, 544. [CrossRef]

6. Toral, A.R.; de los Ríos, A.P.; Hernández, F.J.M.; Janssen, H.A.; Schoevaart, R.; Rantwijk, F.; Sheldon, R.A. Cross-linked *Candida antarctica* lipase B is active in denaturing ionic liquids. *Enzyme Microb. Technol.* **2007**, *40*, 1095–1099.

7. Lee, S.Y.; Vicente, F.A.; Silva, F.A.; Sintra, T.E.; Taha, M.; Khoiroh, I.; Coutinho, J.A.P.; Show, P.L.; Ventura, S.P.M. Evaluating Self-buffering Ionic Liquids for Biotechnological Applications. *ACS Sustain. Chem. Eng.* **2015**, *12*, 3420–3428. [CrossRef]

8. de los Ríos, A.P.; Rantwijk, F.; Sheldon, R.A. Effective resolution of 1-phenyl ethanol by *Candida antarctica* lipase B catalysed acylation with vinyl acetate in protic ionic liquids (PILs). *Green Chem.* **2012**, *14*, 1584–1588. [CrossRef]

9. Hernández, F.J.; de los Ríos, A.P.; Gómez, D.; Rubio, M.; Víllora, G. A new recirculating enzymatic membrane reactor for ester synthesis in ionic liquid/supercritical carbon dioxide biphasic systems. *Appl. Catal. B Environ.* **2006**, *67*, 121–126. [CrossRef]

10. Jiang, Y.; Guo, C.; Xia, H.; Mahmood, I.; Liu, C.; Liu, H. Magnetic nanoparticles supported ionic liquids for lipase immobilization: Enzyme activity in catalyzing esterification. *J. Mol. Catal B Enzym.* **2009**, *58*, 103–109. [CrossRef]

11. Lee, S.H.; Doan, T.T.N.; Ha, S.H.; Koo, Y.M. Using ionic liquids to stabilize lipase within sol–gel derived silica. *J. Mol. Catal B Enzymat.* **2007**, *45*, 57–61. [CrossRef]

12. Zou, B.; Song, C.; Xu, X.; Xia, J.; Huo, S.; Cui, F. Enhancing stabilities of lipase by enzyme aggregate coating immobilized onto ionic liquid modified mesoporous materials. *Appl. Surf. Sci.* **2014**, *311*, 62–67. [CrossRef]

13. Lozano, P.; García-Verdugo, E.; Karbass, N.; Montague, K.; De Diego, T.; Burguete, M.I.; Luis, S.V. Supported Ionic Liquid-Like Phases (SILLPs) for enzymatic processes: Continuous KR and DKR in SILLP–scCO 2 systems. *Green Chem.* **2010**, *12*, 1803–1810. [CrossRef]

14. Suo, H.; Xu, L.; Xu, C.; Chen, H.; Yu, D.; Gao, Z.; Huang, H.; Hu, Y. Enhancement of catalytic performance of porcine pancreatic lipase immobilized on functional ionic liquid modified Fe_3O_4-Chitosan nanocomposites. *Int. J. Biol. Macromol.* **2018**, *119*, 624–632. [CrossRef]

15. Xiang, X.; Suo, H.; Xu, C.; Hu, Y. Covalent immobilization of lipase onto chitosan-mesoporous silica hybrid nanomaterials by carboxyl functionalized ionic liquids as the coupling agent. *Colloid. Surface. B* **2018**, *165*, 262–269. [CrossRef]

16. Nakashima, K.; Kamiya, N.; Koda, D.; Maruyama, T.; Goto, M. Enzyme encapsulation in microparticles composed of polymerized ionic liquids for highly active and reusable biocatalysts. *Org. Biomol. Chem.* **2009**, *7*, 2353–2358. [CrossRef]

17. Campàs, M.; Marty, J.L. *Immobilization of enzymes and cell*, 2nd ed.; Humana Press: Totowa, NJ, USA, 2006; Chapter 7; pp. 77–85. ISBN 1-59745-053-7.

18. Haj-Kacem, S.; Galai, S.; de los Ríos, A.P.; Hernández-Fernandez, F.J.; Smaali, I. New efficient laccase immobilization strategy using ionic liquids for biocatalysis and microbial fuel cells applications. *J. Chem. Technol. Biot.* **2018**, *93*, 174–183. [CrossRef]

19. HajKacem, S.; Galai, S.; Hernández-Fernandez, F.J.; de los Ríos, A.P.; Smaali, I.; Quesada-Medina, J. Bioreactor membranes for laccase immobilization optimized by ionic liquids and cross-linking agents. *Appl. Biochem. Biotechnol.* **2020**, *1*, 1–17. [CrossRef]

20. de los Ríos, A.P.; Hernández-Fernández, F.J.; Martínez, F.A.; Rubio, M.; Víllora, G. The effect of ionic liquid media on activity, selectivity and stability of *Candida antarctica* lipase B in transesterification reactions. *Biocatal. Biotransfor.* **2007**, *25*, 151–156. [CrossRef]

21. Nara, S.J.; Harjani, J.R.; Salunkhe, M.M. Lipase-catalysed transesterification in ionic liquids and organic solvents: A comparative study. *Tetrahedron Lett.* **2002**, *43*, 2979–2982. [CrossRef]

22. van Rantwijk, F.; Secundo, F.; Sheldon, R.A. Structure and activity of *Candida antarctica* lipase B in ionic liquids. *Green Chem.* **2006**, *8*, 282–286. [CrossRef]

23. Lozano, P.; De Diego, T.; Guegan, J.-P.; Vaultier, M.; Iborra, J.L. Stabilization of α-chymotrypsin by ionic liquids in transesterification reactions. *Biotechnol. Bioeng.* **2001**, *75*, 563–569. [CrossRef] [PubMed]

24. Galai, S.; de los Ríos, A.P.; Hernández-Fernández, F.J.; Kacemc, S.; Tomas-Alonso, H.F. Over-activity and stability of laccase using ionic liquids: Screening and application in dye decolorization. *RSC Adv.* **2015**, *5*, 16173–16189. [CrossRef]

25. de Diego, T.; Lozano, P.; Gmouh, S.; Vaultier, M.; Iborra, J.L. Understanding Structure–Stability Relationships of Candida antartica Lipase B in Ionic Liquids. *Biomacromology* **2005**, *6*, 1457–1464. [CrossRef] [PubMed]

26. Hernández-Fernández, F.J.; de los Ríos, A.P.; Gómez, D.; Rubio, M.; Víllora, G. Enhancement of activity and selectivity in lipase-catalyzed transesterification in ionic liquids by the use of additives. *J. Chem. Technol. Biotechnol.* **2007**, *82*, 882–887. [CrossRef]

27. de los Ríos, A.P.; Hernández-Fernández, F.J.; Tomás, P.; Gómez, D.; Víllora, G. Synthesis of esters in ionic liquids. The effect of vinyl esters and alcohols. *Process Biochem.* **2008**, *43*, 892–895. [CrossRef]

28. Hernández-Fernández, F.J.; de los Ríos, A.P.; Lozano-Blanco, L.J.; Godínez, C. Biocatalytic ester synthesis in ionic liquid media. *J. Chem. Technol. Biot.* **2010**, *85*, 1423–1435. [CrossRef]

29. Machado, M.C.; Webster, T.J. Lipase degradation of plasticized polyvinyl chloride endotracheal tube surfaces to create nanoscale features. *Int. J. Nanomedicine* **2017**, *12*, 2109–2115. [CrossRef]

30. Sheldon, R.A. Cross-linked enzyme aggregates (CLEA®): Stable and recyclable biocatalysts. *Biochem. Soc. Trans.* **2007**, *35*, 1583–1587. [CrossRef]

31. Sheldon, R.A. The E factor 25 years on: The rise of green chemistry and sustainability. *Green Chem.* **2017**, *19*, 18–43. [CrossRef]

32. Tobiszewski, M.; Marć, M.; Gałuszka, A.; Namieśnik, J. Green chemistry metrics with special reference to green analytical chemistry. *Molecules* **2015**, *20*, 10928–10946. [CrossRef] [PubMed]

33. Lozano, P.; De Diego, T.; Sauer, T.; Vaultier, M.; Gmouh, S.; Iborra, J.L. On the importance of the supporting material for activity of immobilized *Candida antarctica* lipase B in ionic liquid/hexane and ionic liquid/supercritical carbon dioxide biphasic media. *J. Supercrit. Fluids* **2005**, *40*, 93–100. [CrossRef]

Sample Availability: Samples of the enzyme derivatives are available from the authors.

 *molecules*

Article

Bio-Based Solvents and Gasoline Components from Renewable 2,3-Butanediol and 1,2-Propanediol: Synthesis and Characterization

Vadim Samoilov [1,*], Denis Ni [1], Arina Goncharova [1], Danil Zarezin [1], Mariia Kniazeva [1], Anton Ladesov [2], Dmitry Kosyakov [2], Maxim Bermeshev [1] and Anton Maximov [1]

[1] A.V. Topchiev Institute of Petrochemical Synthesis, Russian Academy of Sciences (TIPS RAS), 29 Leninsky Prospect, 119991 Moscow, Russia; max@ips.ac.ru (A.M.)

[2] Core Facility Center "Arktika", Northern (Arctic) Federal University, 17 Severnaya Dvina Embankment, 163002 Arkhangelsk, Russia; lokoal13@gmail.com (A.L.); d.kosyakov@narfu.ru (D.K.)

* Correspondence: samoilov@ips.ac.ru

Received: 4 March 2020; Accepted: 1 April 2020; Published: 9 April 2020

Abstract: In this study approaches for chemical conversions of the renewable compounds 1,2-propanediol (1,2-PD) and 2,3-butanediol (2,3-BD) that yield the corresponding cyclic ketals and glycol ethers have been investigated experimentally. The characterization of the obtained products as potential green solvents and gasoline components is discussed. Cyclic ketals have been obtained by the direct reaction of the diols with lower aliphatic ketones (1,2-PD + acetone → 2,2,4-trimethyl-1,3-dioxolane (TMD) and 2,3-BD + butanone-2 → 2-ethyl-2,4,5-trimethyl-1,3-dioxolane (ETMD)), for which the ΔH^0_r, ΔS^0_r and ΔG^0_r values have been estimated experimentally. The monoethers of diols could be obtained through either hydrogenolysis of the pure ketals or from the ketone and the diol via reductive alkylation. In the both reactions, the cyclic ketals (TMD and ETMD) have been hydrogenated in nearly quantitative yields to the corresponding isopropoxypropanols (IPP) and 3-sec-butoxy-2-butanol (SBB) under mild conditions (T = 120–140 °C, $p(H_2)$ = 40 bar) with high selectivity (>93%). Four products (TMD, ETMD, IPP and SBB) have been characterized as far as their physical properties are concerned (density, melting/boiling points, viscosity, calorific value, evaporation rate, Antoine equation coefficients), as well as their solvent ones (Kamlet-Taft solvatochromic parameters, miscibility, and polymer solubilization). In the investigation of gasoline blending properties, TMD, ETMD, IPP and SBB have shown remarkable antiknock performance with blending antiknock indices of 95.2, 92.7, 99.2 and 99.7 points, respectively.

Keywords: renewable solvents; ketals; ethers; 2,3-butanediol; renewable fuel; propylene glycol; Kamlet-Taft

---

1. Introduction

The use of fossil fuel feedstocks for the production of energy carriers and chemicals has been developed a lot during the XXth century, and can now be considered as one of the main sources of current ecological problems. Among the latter there are environmental pollution (particularly, air, water and soil contamination) and climate change, which influence each other. One of the possible solutions to the aforementioned problems could be the extensive development of the biomass processing industry that provides the chance to replace fossil-based fuels and chemicals with bio-based renewable analogues.

Renewable glycols such as 1,2-propanediol (1,2-PD) and 2,3-butanediol (2,3-BD) are of great interest as potential starting materials for chemical conversions for several reasons. First, they can be readily obtained from biomass sources by different methods: 1,2-PD can be obtained either by bioglycerol hydrogenolysis [1,2] or by microbial production starting from carbohydrates [3]. The process of carbohydrate biomass fermentation, which yields 2,3-BD, is remarkable as it provides a relatively high diol yield as well as overall productivity along with a quite low energy consumption. The low host toxicity of 2,3-BD which is conducive to a higher possible product titer in the fermentation broths should be also mentioned [4–7]. The direct microbial conversion route of CO_2 into 2,3-BD is also worth mentioning [8]. Hence, the noted compounds are of particular interest as renewable feedstocks for the production of chemicals.

The majority of studies on 2,3-BD conversion are dedicated to its dehydrative conversion that yields methyl ethyl ketone (MEK) and isobutyraldehyde [9–12], 3-butene-2-ol [10,13–15], butenes [16–18] and 1,3-butadiene [19–21]. Some 2,3-BD derivatives have been previously considered as potential motor fuel components and organic solvents. As far as this aspect is concerned, one should note the recent work by Harvey [22], which has arisen interest to the one-step 2,3-BD dehydrative conversion into a cyclic ketal by condensation with MEK generated in situ (Scheme 1), previously developed by Neish et al. in 1945 [23].

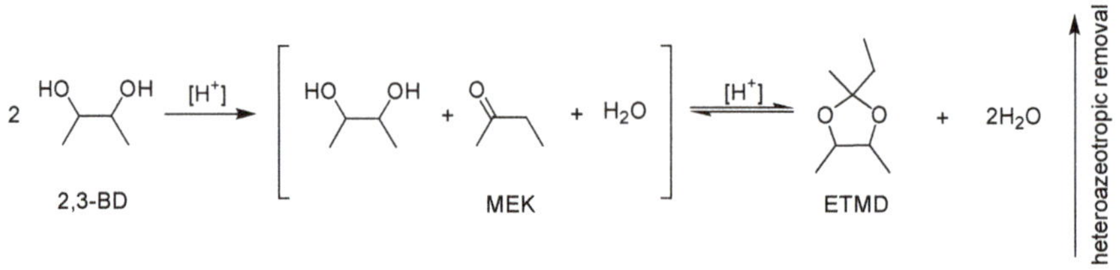

Scheme 1. One step-conversion of 2,3-BD into a cyclic ketal.

The main feature of the abovementioned approach to 2,3-BD conversion is the continuous stripping of the reaction products (water and the cyclic ketal) out of the reaction mixtures, which may provide favorable conditions for driving the reversible ketalization reaction to completion. The one-step conversion might also be partially combined with the drying of 2,3-BD, since some water content in the feed diol does not decrease the target product yield. Nearly quantitative yields of the ketal could be obtained in this process by applying a proper combination of process conditions such as temperature, pressure and acid catalyst concentration [23].

The cyclic ketal derived from 2,3-BD — 2-ethyl-2,4,5-trimethyl-1,3-dioxolane — has been characterized as a gasoline component and an organic solvent by Harvey et al. [22]. As a gasoline additive, the compound is significant for its good miscibility with fuel, high lipophilicity, and acceptable boiling point (132–142 °C depending on the isomeric composition), relatively high antiknock performance (RON/MON = 93.5/86.7) and calorific value (28.3 MJ L^{-1} versus 21.1 MJ L^{-1} for ethanol). Although the use of ETMD as a renewable solvent has been proposed, the existing data is likely insufficient for understanding its potential. In addition, in the aforementioned study by Harvey et al., some properties of the pure compound including the neat octane numbers were reported without the blending octane numbers and without any detailed data on the base gasoline properties. Some data on the diesel fuel blending properties of 2,3-BD acetals had also been reported earlier by Staples et al. [24]. Thus, a deeper investigation of both solvent and gasoline-blending properties of the 2,3-BD ketal derivatives is reasonable for exploring its real application potential.

Renewable 1,2-PD could also be transformed into a cyclic ketal via traditional acid-catalyzed ketalization with acetone [25]. It should also be noted that the formation of TMD in the vapor-phase hydrogenolysis of solketal has been recently reported [26]. Some data on 1,2-PD cyclic ketals regarding their diesel fuel blending and physical/solvent properties can be found in the literature [25,27], whereas no data on the gasoline blending properties could be found by us.

The combination of the ketalization process with the non-destructive ketal hydrogenolysis reaction that yields the corresponding ethers (Scheme 2) offers intriguing prospects.

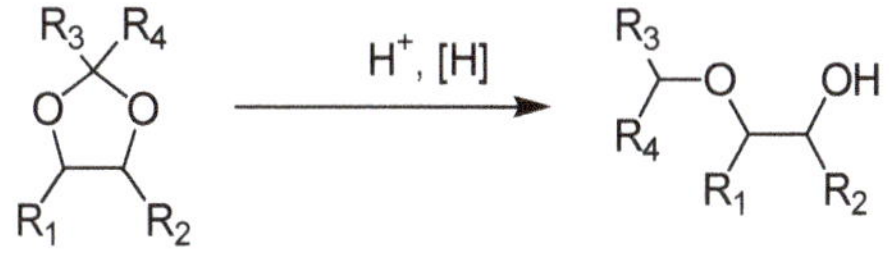

Scheme 2. Synthesis of the glycol ethers by hydrogenolysis of the corresponding cyclic ketals.

This reaction can be conducted with various reductive agents such as $LiAlH_4$-$AlCl_3$ [28–31], $LiAlH_4$-BF_3 [32], $NaBH_4$ [33,34], TMDS [35,36] and hydrogen [37–40]. In the latter case a bifunctional catalyst system is required, consisting of either a heterogeneous Pd/C hydrogenation catalyst in combination with a homogeneous acid (*p*-TSA or camphorsulfonic acid) or one based on palladium supported on an acidic aluminosilicate or an acidic carbon [41,42]. When employing hydrogen as a reducing agent (what can be considered the most feasible route) nearly quantitative ether yields could be achieved. In this manner, the renewable diols could be converted into the corresponding cyclic ketals and then they could be subjected to selective hydrogenolysis that yields renewable glycol ethers. The compounds of the latter group are the well-known organic solvents and hydrotropes with multiple applications [43,44]. A variation of the approach of interest is the direct reductive alkylation reaction between a diol and a carbonyl compound. In this case, the synthesis of the glycol ether can be conducted in a one-step process under mild conditions (e.g., under moderate hydrogen pressure of 1–4 MPa and 100–140 °C) [38,40]. A number of recent papers were dedicated to the conversion of renewable diols, namely, 1,2-propanediol and glycerol, into ethers and ketals, as well as to the characterization of the mentioned derivatives as potential renewable fuel components [45] and solvents [27,43,46]. Thus, the development of approaches for the synthesis of bio-based glycol ethers might be of interest for the renewable petrochemical substituents production.

The purposes of the study reported herein are: (a) to investigate the regulations of the synthesis of cyclic ketals from diols (1,2-PD and 2,3-BD) and ketones (acetone and MEK); (b) to describe and to perform some experimental evaluations of approaches to glycol ethers synthesis either by cyclic ketal hydrogenolysis or via reductive alkylation; (c) to characterize the cyclic ketal and glycol ether derivatives regarding their potential applicability as organic solvents and gasoline components; and (d) to perform a primary sustainability evaluation of the proposed routes towards bio-based solvents.

2. Results and Discussion

2.1. The Ketalization of 1,2-PD and 2,3-BD with Acetone and MEK

The ketalization between diols and acetone (or MEK) should comply with the general rules known for the reactions of this type. As is known, the ketalization process is slightly exothermic and results in a decrease in entropy [47]. Thus, to obtain the maximum ketal yields lower temperatures are preferred. At the same time, the thermodynamic stability of the cyclic ketal products depends on the molecular structure of the precursor diols and carbonyl compounds, what is supported by the different equilibrium yields obtained either within the reactions of the diol with the different carbonyl compounds [48–50] or in transacetalization (transketalization) reactions [51]. In order to determine the thermodynamic equilibrium for the reactions of interest (yielding the corresponding cyclic acetals from 1,2-PD and 2,3-BD) and to evaluate the relation between the reactants molecular structure and the reaction thermodynamic, experimental measurements of the equilibrium constant temperature dependence were conducted.

The determination of equilibrium compositions was carried out under the conditions employed earlier for the ketalization between acetone and glycerol [52,53]. During the analysis of the reaction mixtures, no byproducts were observed. The experimental results obtained (Table 1) show the dependence between the equilibrium yield and the temperature typical for the homologous ketalization reactions. The data was plotted in the Arrhenius coordinates (Figure 1) and fitted with a linear function with good precision.

Table 1. The equilibrium compositions of the reaction mixtures obtained by the ketalization of 1,2-PD with acetone and 2,3-BD with MEK. The conditions: the ambient air pressure, keton:diol = 6:1 mol, 5 wt. % (to diol) of Amberlyst 36 dry as the catalyst.

T, K	x^0_{ketone}	x^0_{diol}	x^1_{ketone}	x^1_{diol}	x^1_{ketal}	x^1_W	K_c	X_{eq}
				1,2-PD + acetone				
298	0.770	0.115	0.687	0.012	0.093	0.093	1.058	0.897
303	0.770	0.115	0.681	0.015	0.095	0.095	0.902	0.874
313	0.770	0.115	0.697	0.015	0.087	0.087	0.705	0.867
323	0.770	0.115	0.689	0.019	0.089	0.089	0.593	0.833
				2,3-BD + MEK				
298	0.540	0.090	0.014	0.471	0.073	0.073	0.832	0.849
303	0.540	0.090	0.014	0.472	0.072	0.072	0.753	0.839
313	0.540	0.090	0.017	0.475	0.069	0.069	0.591	0.812
323	0.540	0.090	0.018	0.476	0.068	0.068	0.524	0.796
333	0.540	0.090	0.022	0.479	0.065	0.065	0.402	0.756

[1] The X_{eq} data shown as the mean have been derived from the six samples ± SD (0.005 and 0.002 for 1,2-PD and 2,3-BD ketalization, respectively). x^0_{ketone}, x^0_{diol} are the initial molar fractions of the ketal and the diol, respectively; x^1_{ketone}, x^1_{diol}, x^0_{ketal}, x^0_W are the final (at the equilibrium state) molar fractions of the ketone, the diol, the ketal and water, respectively; K_c is the equilibrium constant; X_{eq} is the diol equilibrium conversion.

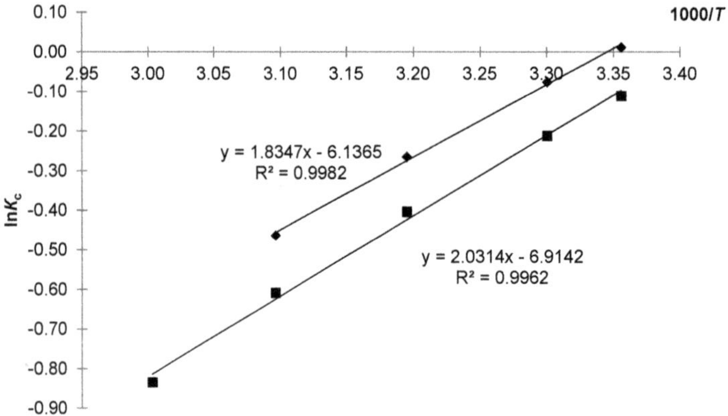

Figure 1. The influence of the reaction temperature on the equilibrium constant in the reactions of 2,3-BD with MEK (■) and in the reactions of 1,2-PD with acetone (♦).

Using the Van 't Hoff equation in the form:

$$lnK_c = \frac{\Delta S^0}{R} - \frac{\Delta H^0}{R}\frac{1}{T} \qquad (1)$$

and the equation which expresses the relation between the changes Gibbs free energy, the enthalpy and the entropy upon the reaction:

$$\Delta G^0 = \Delta H^0 - T\Delta S^0 \qquad (2)$$

the corresponding values of ΔG^0, ΔH^0 and ΔS^0 for the reaction were calculated (Table 2).

Table 2. The values of thermodynamic parameters for the diol ketalization reactions.

Reaction	ΔH^0_r, kJ mol^{-1}	ΔS^0_r, J mol^{-1} K^{-1}	ΔG^0_r, kJ mol^{-1}	Reference
2,3-BD + MEK	-16.6 ± 1.0	-56.8 ± 3.3	0.30 ± 0.02	this study
2,3-BD + acetone [1]	-16.7	-56.5	0.5	[47]
1,2-PD + acetone	-15.1 ± 0.9	-51.1 ± 3.1	0.05 ± 0.02	this study
1,2-PD + acetone [1]	-3.3	-17.6	1.7	[47]
glycerol + acetone	-30.1 ± 1.6	-100.0 ± 10.0	-2.1 ± 0.1	[52]
glycerol + acetone	-14.5	-49.9	0.4	[54]
glycerol + acetone	-19.8 ± 1.6	-64.4 ± 5.2	-0.6 ± 0.05	[53]
glycerol + acetone [1]	-15.5	-54.4	0.7	[47]
GMME [2] + acetone	-15.8 ± 1.2	-51.5 ± 3.9	-0.4 ± 0.03	[53]

[1] The values for T = 300 K, the correlation coefficient r = 0.975–0.998; [2] No uncertainty data has been reported; [3] GMME = glycerol 1-monomethyl ether. The data has been shown as the mean ± SD.

A comparison between the values obtained in this work with the previously reported ones permits us to conclude that in the present case the differences in the molecular structure do not affect the thermodynamic stability of the cyclic ketals formed. The only result which shows a significant difference seems to be the result of Nanda et al. [52] for the reaction of glycerol with acetone, probably due to the use of ethyl alcohol as solvent, while other results were obtained in solventless reactions. The values of ΔH^0_r and ΔS^0_r obtained for the reaction between 1,2-PD and acetone by Anteunis and Rommelaere are likely to be underestimated, what might be connected with the known precision limits of the NMR measurements. At the same time, the data from the Table 2 proves the postulate of Anteunis and Rommelaere about the isoequilibrium relationship for the homological reactions of the cyclic ketal formation. For those typical values of ΔH^0_r, ΔS^0_r are about -17.2 ± 2.6 kJ mol^{-1} and 57.2 ± 7.3 J mol^{-1} K^{-1}, respectively.

The equilibrium yield values for the reactions of 2,3-BD and 1,2-PD ketalization are close to the values measured for glycerol ketalization under the similar conditions [52,53]. Pure TMD and ETMD samples were obtained via the direct ketalization under the same ketone molar excess (6:1) conditions. The isolated yields of TMD and ETMD (83.7 and 81.6%) turned out to be close to the equilibrium yields determined by GC (89.7 and 84.9%), hence the isolated yields for TMD and ETMD amounted to 0.933 and 0.961 of the corresponding equilibrium yields. Thus, the direct synthesis approach offers excellent yields along with a relatively simple set-up, and thus might be further potentially developed with a view to an industrial process. The thermodynamic data obtained here could be of interest for the necessary reaction engineering purposes.

2.2. The Glycol Ethers Synthesis Via the Hydrogenolysis of the Corresponding Cyclic Ketals

The properties of the catalysts used for 1,2-PD and 2,3-BD monoether synthesis are given in Table 3. The performance of Pd/Al-HMS materials in the cyclic glycerol ketal hydrogenolysis reaction and the characterization details have been previously reported [39], so in this paper only the brief description is given. The values of the specific surface areas for the catalysts indicate well-developed pore structures (between 680 m^2 g^{-1} and 850 m^2 g^{-1}) and a narrow pore size distribution with mean pore sizes of about 3–5 nm. The acidity of the mesoporous aluminosilicas demonstrates a non-linear growth with the increase in the Al content. The acidity strength distribution pattern for Al-HMS aluminosilicas determined previously has shown that strong acid sites are practically absent, what might provide the high selectivity in the reaction of the cyclic ketals hydrogenolysis [39].

The palladium particles are well dispersed, with mean sizes between 3 nm and 5 nm, what allows us to reach the conclusion that the metal particles size seems to be limited by the pore size and that the metal particles are predominately located inside the pores of the Al-HMS. The metal particles dispersion degrees in the catalysts with the supports of the different Si/Al ratio are practically identical.

Table 3. The physicochemical properties of the Al-HMS-supported catalysts.

Support	SiO_2/Al_2O_3	S_{BET}, m^2 g^{-1}	V_{total}, cm^3 g^{-1}	Total Concentration of Acid Sites, mmol g^{-1}	Pd Particles Mean Size, nm [1]
Al-HMS(10)	9.8	680	0.7	256	3.8
Al-HMS(15)	14.7	770	0.9	211	4.2
Al-HMS(20)	19.7	850	1.0	170	3.5
Al-HMS(30)	29.6	820	1.1	144	4.1

[1] The values for 2 wt.% Pd/Al-HMS supported catalysts have been calculated according to the TEM results; S_{BET} is the BET specific surface area; V_{total} is the total pore volume.

The catalysts were tested in the cyclic ketal (both TMD and ETMD) hydrogenolysis reaction yielding the corresponding glycol ethers (Scheme 3). The reaction is known for its high selectivity: the yields are usually nearly quantitative for the reduction with hydrogen [38,41], LiAlH$_4$ [55] or 1,1,3,3-tetramethyldisiloxane [36]. The results of the catalytic tests (Table 4) demonstrate that in the present case the selectivity values were also very high, normally reaching 97–98 mol %. The corresponding glycol ether (IPP for TMD, SBB for ETMD) was actually the main and only reaction product with the remaining 2–3% of selectivity accounted for by the hydrolysis products (the ketone and the diol), originated from the trace water.

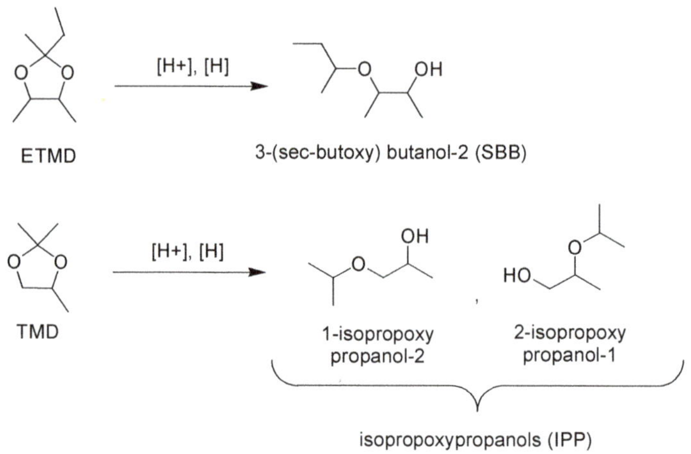

Scheme 3. The hydrogenolysis of the cyclic ketals (the reaction scheme).

The hydrogenolysis of the glycerol-acetone cyclic ketal (solketal) over these catalysts has been studied earlier. It has been revealed by the authors that the Si/Al ratio of the support influences the overall catalytic activity [39]. The observed differences were attributed to the different activity of the supports in the acid-catalyzed reactions, e.g., ketalization, as more hydrophobic supports with greater Si/Al ratio showed to be more active. The strong adsorption of polar species (for example, water, polar organic compounds), which are typical for the relatively hydrophilic materials, might be the reason for their limited catalytic activity [50,52]. Thus, in the case of solketal catalytic hydrogenolysis, more hydrophilic material is likely to adsorb the primary hydrogenolysis product that is the diol.

In the present case, the authors have not observed any significant differences between the palladium catalysts supported over the Al-HMS materials with the Si/Al ratios ranging from 10 to 30 (Table 4). The Al-HMS(10) catalyst was slightly less active at the lower temperature, while the activities of all the catalysts under $T = 160\ °C$ were practically of the same value. This lets one suppose that in the present case there is no significant correlation between the support hydrophobicity and the catalytic activity, since the difference in the polarity of the reactant and the product (the ketal and the corresponding ether) is likely to be relatively low. As shown below, the physical properties in the "diol ketal–glycol monoether" pairs (for example, the water miscibility, boiling points and vapor pressures) are much closer compared to the "solketal–glycerol monoisopropyl ether" pair.

By conducting the reaction for more time, a nearly quantitative yield of the ether might be obtained. Under the optimized conditions (Table 4, entry 5) both the TMD and ETMD were converted into the corresponding ethers with yields over 90 mol % (91.4 and 92.1 mol %, respectively), thus indicating at the feasibility of the approach for the catalytic synthesis of the renewable glycol ethers.

Table 4. The hydrogenolysis of TMD and ETMD over 2% wt. Pd/Al-HMS catalysts with the different supports. The conditions are: $p(H_2) = 40$ bar, 100 mg of the catalyst, 2.5 mL of the ketal, the reaction time 5 h.

Entry No.	Support	T, °C	Y_{IPP}, mol % [1]	Y_{SBB}, mol % [1]
1	Al-HMS(10)	140	29.8	11.0
2		160	40.5	30.6
3	Al-HMS(15)	140	32.7	31.7
4		160	44.2	51.1
5 [2]		160	91.4	92.1
6	Al-HMS(20)	140	38.6	33.3
7		160	53.2	35.6
8	Al-HMS(30)	140	19.9	31.3
9		160	13.9	25.3

[1] Ketal to ether hydrogenolysis selectivity 97–98 mol % in all the cases; [2] reaction time, 24 h; Y_{IPP}, Y_{SBB}—the GC yields of IPP and SBB, respectively.

For the synthesis of ethers via the catalytic hydrogenolysis of the corresponding acetals and ketals high regioselectivity values are typical [39,40]. While the hydrogenolysis of ETMD yields only one structural isomer, the hydrogenolysis of TMD could yield two isomeric 1,2-PD monoethers: 2-isopropoxy-1-propanol and 1-isopropoxy-2-propanol. According to our results, the regioselectivity for 1-isopropoxy-2-propanol reached 87–91%. This value is lower than the previously observed one in solketal hydrogenolysis and it is closer to the regioselectivity of the solketal *O*-isopropyl ether reduction [39]. Close regioselectivity values have been also reported for the direct reductive alkylation of glycerol with aliphatic aldehydes [41]. In the preparative syntheses carried out in the given work upon the reduction of TMD in the $LiAlH_4$-$AlCl_3$ system the regioselectivity for the 1-isopropoxy-2-propanol was about 82%, being here typical for this approach [29].

Since ETMD might be obtained from 2,3-BD with the excellent yield in the one-step process, this compound seems to be the more preferable one as the starting material for the synthesis of the corresponding glycol ether by the pure ketal hydrogenolysis comparing with 2,3-BD. The renewable 1,2-PD may be obtained by the catalytic hydrogenolysis of bioglycerol. 1,2-PD cost is comparable to the price of acetone, so the feasibility of the one-step conversion of 1,2-PD to TMD in the manner that is similar to 2,3-BD conversion to ETMD seems to be controversial. Thus, the conversion of 1,2-PD into the corresponding glycol ether by reductive alkylation with acetone (the latter can originate from sugars by the ABE fermentation) seems to be of value (Scheme 4).

Scheme 4. The reductive alkylation of acetone with 1,2-PD over a bifunctional catalyst.

The reductive alkylation reaction is generally believed to run through the formation of the intermediate ketal followed by its catalytic hydrogenation [38]. Therefore, the influence of the reaction temperature should be crucial: its increase enhances the hydrogenolysis rate, while decreasing the equilibrium concentration of the intermediate ketal described above in the ketalization thermodynamic study. The experimental results (Table 5) have shown that there is a temperature optimum for the reductive alkylation reaction. For example, upon the reductive alkylation of 1,2-PD with acetone under T = 120, 140, 160 °C (Table 5, entries 7–9), the yields of IPP amounted to 15.2, 58.7 and 7.9%, respectively. Hence, at the lower temperature the ether yield decreased along with the hydrogenation rate. At 160 °C the obtained ether yield was connected with the unfavorable influence of the temperature on the equilibrium between diol and the corresponding ketal, which is the reaction intermediate. The decrease of the diol:ketone ratio from 40:1 to 20:1 (mol/mol) (Table 5, the entries 7 and 12) also led to an ether yield decrease (from 15.2 to 6.8%), thus supporting the determinant influence of the ketal equilibrium concentration on the target product yield. Regarding the reaction with the greater reaction time, the glycol ether yields are likely to be as high as 92.6 and 93.3 mol % for IPP and SBB, respectively (Table 5, entries 5 and 11). The regioselectivity between the IPP structure isomers was close to the values obtained in the direct TMD hydrogenolysis (about 90–91% to 1-isopropoxy-2-propanol).

Table 5. The reductive alkylation of 1,2-PD and 2,3-BD with acetone and MEK with 2%Pd/Al-HMS(20) bifunctional catalyst. $p(H_2)$ = 40 bar, 10 mg catalyst, 2.5 mL of a diol.

Entry No.	Diol:Ketone Molar Ratio	T, °C	t, h	X_{ketone}, %	Y_{ketal}, %	Y_{ether}, %	$Y_{alcohol}$, %
MEK + 2,3-BD							
1	40	120	5	73.7	56.1	16.9	0.7
2	40	140	5	97.1	34.4	62.2	0.6
3	40	160	5	62.2	51.6	9.6	1.0
4	40	140	10	98.3	17.7	79.9	0.7
5	40	140	20	98.8	5.3	92.6	0.9
6	20	120	5	55.1	47.9	6.2	1.0
Ac + 1,2-PD							
7	40	120	5	86.0	70.2	15.2	0.5
8	40	140	5	96.3	36.5	58.7	1.1
9	40	160	5	82.4	71.1	7.9	3.4
10	40	140	10	97.8	15.4	81.3	1.1
11	40	140	20	98.0	3.6	93.3	1.1
12	20	120	5	85.0	77.2	6.8	1.0

X_{ketone}—conversion of the ketone (MEK/acetone), Y_{ketal}, Y_{ether}, $Y_{alcohol}$—the GC yields of the ketal (ETMD/TMD), the ether (SBB/IPP) and the alcohol (2-butanol/isopropanol), respectively.

The reusability of the bifunctional catalyst was checked in both the pure ETMD hydrogenolysis and 2,3-BD + MEK reductive alkylation reactions (Table 6). After the first ketal hydrogenolysis cycle a slight increase in the catalyst activity was observed (the SBB yield increased from 35.6 to 36.9 mol %), probably, due to the pre-reduction of the catalyst. For the subsequent cycles, a slight decrease in activity was observed: in the fifth cycle, the target product yield amounted to 92.5% of the initial value. In the case of the 2,3-BD reductive etherification with MEK, after the first cycle the target product yield decreased sharply from 62.6 to 54.7 mol %. In the next cycles, the activity change was lower, despite the fact that the SBB yield gradually decreased to 52.1 mol % (83.8% of the initial value). This pattern of the activity change observed in the latter reaction might be attributed to the adsorption of water on the catalyst active sites: the fresh catalyst sample showed the higher activity as it was dry, while the spent catalyst contacted with the water that was released in the catalytic reaction. It should be taken into the account that the presence of water might influence the reductive etherification by the affecting on the ketalization equilibrium [39] or the palladium reduction process [42].

Table 6. The synthesis of SBB over 2% wt. Pd/Al-HMS (15): the catalyst reusability test. The conditions are: p(H$_2$) = 40 bar, 100 mg of the catalyst, 2.5 mL of the ketal, the reaction time was 5 h.

Cycle No.	Y_{SBB}, mol %	
	ETMD Hydrogenolysis [1]	**2,3-BD Reductive Alkylation**
1	35.6	62.2
2	36.9	54.7
3	35.3	53.3
4	35.1	53.1
5	32.5	52.1

[1] Ketal to ether hydrogenolysis selectivity 97–98 mol % in all the cases; Y_{IPP}, Y_{SBB}—the GC yields of SBB.

Based on the results obtained, these approaches, namely the hydrogenolysis of the pure ketal obtained in the separate synthesis from the diol and the ketone, and the one-step reductive ketalization alkylation, both turned out to be feasible for the synthesis of the renewable glycol ethers. The use of bifunctional Pd/Al-HMS catalysts permits the reactions to be carried out under relatively mild conditions (T = 120–140 °C) with excellent selectivities (97–98 mol % and 94–95 mol % for the ketal hydrogenolysis and the reductive ketalization alkylation) and in nearly quantitative yields. The recovery of the target product from the reaction mixtures is likely to be easily conducted by simple distillation.

2.3. The Characterization of the Products

The search of the new bio-based organic solvents for the substitution of the petrochemical-derived ones is an important problem of sustainable chemistry. Investigations of solvent properties of compounds derived from propylene glycol, glycerol, levoglucosane and isosorbide are to be mentioned herein as the examples [27,43,44,46,56,57]. In the given study, the efforts have been taken to describe the properties of the synthesized compounds, namely, of two cyclic ketals—TMD and ETMD and of two corresponding glycol ethers—IPP and SBB. The estimate of the properties which might be relevant for the organic solvents was performed with the respect to the criteria for the green solvents recognizing reported earlier by Jessop [58].

All the compounds tested appear to be the low-viscosity liquids with low melting points (<−60 °C) and the pleasant fruity smell. The value of the TMD viscosity is in good accordance with the data reported earlier by Kapkowski et al. [25]; it might be compared to the kinematic viscosity of *n*-butyl acetate (0.78 mm^2 s^{-1}), while the viscosity of ETMD is of the same order as that of diethyl ether (0.31 mm^2 s^{-1}). The viscosities of IPP and SBB are close to the value of 2-butoxyethanol (3.64 mm^2 s^{-1}). The boiling points values (Table 7) being less than 250 °C under atmospheric pressure place these compounds in the VOC group according to the EU classification.

Table 7. The main physicochemical properties of the ketals and the corresponding ethers.

Compound	bp, °C	mp, °C	d_{20}, kg cm^{-3}	n_D^{20}	v_{20}, mm^2 s^{-1}	η_{20}, cP [1]	NHOC, kJ/kg	NHOC, MJ/L
TMD	98–99	<−60	0.900	1.3940	0.75	0.68	28553 ± 43	25.70
ETMD	140–142	<−60	0.899	1.4110	0.23	0.20	31556 ± 12	28.37
IPP	144–145	<−60	0.883	1.4100	3.01	2.66	29839 ± 10	26.35
SBB	163–164	<−60	0.875	1.4175	3.89	3.40	32548 ± 5	28.48

[1] The parameter was calculated from the experimentally measured density and the kinematic viscosity.

For volatile organic compounds not only the boiling points, but also the saturated vapor pressure and the evaporation rate values are important. In the given study, the evaporation rates were estimated according to the thermogravimetric method, whose applicability was demonstrated earlier by the Aubry group [59]. The results (Figure 2, Table 8) demonstrate that TMD evaporates at the greater rate than the reference BuOAc (the evaporation rate is 1.30), while other compounds evaporate slightly slower (ETMD – 0.91, IPP – 0.84 and SBB – 0.73). The evaporation rates of the ketals might be expected to be higher than the values obtained for the corresponding ethers.

In order to characterize the volatility, the temperature dependences of saturated vapor pressure have been determined experimentally (Table S1). The coefficients of the Antoine equation have been calculated (Table 9) by means of the mathematical regression of the experimental results. It has been reported for the case of glycol and glycerol monoalkyl ethers that the TGA-derived evaporation rates correlate with the saturated vapor pressure values, but not with the boiling points [59]. In the given study the linear correlations have been observed between both the boiling point–evaporation rate (RSD = 0.984) and the saturated vapor pressure–evaporation rate (RSD = 0.977) (Figures S1 and S2). One should note that the results of the TGA evaporation rate measurements should be employed only for the primary qualitative estimate: for the isomer of IPP with the close boiling point (propylene glycol *n*-propyl ether C_3P_1, bp = 149 °C), a RER of 0.56 was reported [59], which is about a third lower than the value obtained by us. Moreover, the reported RER value for C_3P_1 showed to be about 23% lower than the one obtained for SBB, which has the highest boiling point among the compounds tested. At the same time, the experimental SVP values (at 50 °C) for IPP (1956 Pa) and its isomer, propylene glycol *n*-propyl ether 1756 Pa) [59], are in the proper relation.

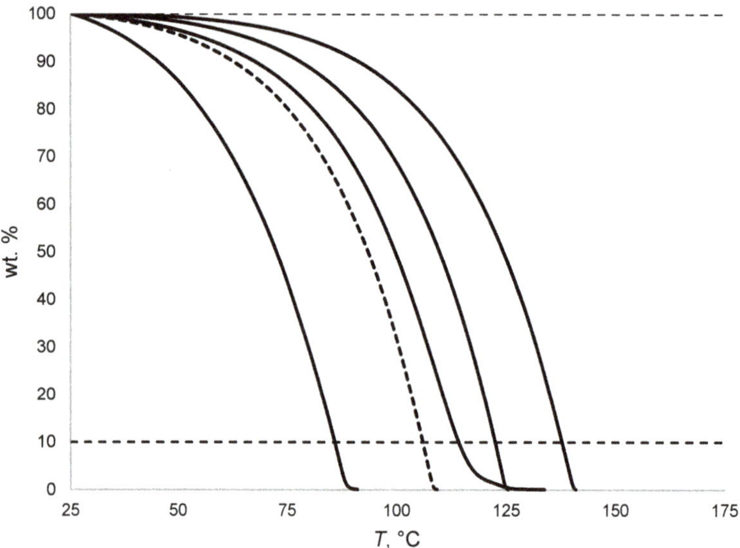

Figure 2. The TGA evaporation curves for IPB and SBB. From left to the right: TMD, BuOAc, ETMD, IPP, SBB.

Table 8. The evaporation rates for the renewable diols derivatives calculated from the TGA measurements.

Compound	TMD	ETMD	IPP	SBB
$T_{10\%}$, °C	85.8	114.4	122.3	138.0
Evaporation rate, rel. to BuOAc	1.30	0.91	0.84	0.73

Table 9. The Antoine equation parameters for the 1,2-PD and 2,3-BD derived ketals and ethers.

Compound	Temperatures Range, K	A	B	C
TMD	298–334	3.9164	1280.14	−53.52
ETMD	318–372	4.1551	1618.36	−37.76
IPP	318–372	4.1812	1268.21	−108.16
SBB	324–386	5.2285	2039.98	−47.39

All the compounds tested appeared to be completely miscible with methanol, ethanol, isopropyl alcohol, diethyl ether, toluene and dodecane. Except for IPP, the compounds were slightly miscible with water (Table 10): the miscibility with water decreased with the increase in alkyl substituents molecular weight.

Table 10. The data on the water solubility of the renewable diol derivatives.

Equilibrium Solubility, wt. %	TMD	ETMD	IPP	SBB
Water in compound	1.8	0.3	∞	0.6
Compound in water	3.3	0.9	∞	3.8
LogK_{OW}	1.46	2.51	0.40	1.24

The miscibility with water measured for the ketals turned out to be in all the cases lower than that one of the corresponding glycol ethers. IPP being miscible with water is likely to be considered as the component of the low-melting water-based liquids, that is why for this compound the properties of aqueous solutions have been estimated (Table 11).

Table 11. The IPP aqueous solutions densities and the melting points.

IPP Concentration, wt. %	d_{20}, g/cm^3	mp, °C
100	0.883	<−60
80	0.925	−31.2
60	0.947	−22.5
40	0.959	−20.9
20	0.970	−14.7

One should note that if a liquid with the freezing point of −15 °C is needed, the IPP-based aqueous solution might contain less organic matter (22 vol. %) than in case of ethylene glycol and propylene glycol (both 30 vol. % for −14 °C), glycerol and isopropanol (both 40 vol. % for −15 °C). At the same time, if a lower freezing temperature is required, the only advantage of IPP over propylene glycol is a lower solution viscosity; the lower flash point of IPP compared to 1,2-PD should be considered in this case disadvantageous.

One of the major areas of organic solvents usage is the dissolution of polymers as the search for the new greener solvents for the preparation of the polymer solutions, e.g., for dyes and coatings applications is of great interest. The ketals, TMD and ETMD, turned out to be the excellent solvents for PS and polybutadiene (Table 12): 300 mg of the polymer per 1 mL of the solvent were completely dissolved to give a clear solution, and only the high viscosities of the solutions hindered a further increase in the test polymer concentration. The same polymers appeared to be just slightly soluble in the glycol ethers, which bear OH-groups in the molecular structure. The chlorinated poly(vinylchloride) (CPVC) sample, which has the high solubility in dichloroethane, might dissolve in the ketals and the glycol ethers in quite the low concentrations thus making our hopes that these compounds might partially substitute chlorinated organic solvents for the dissolution of chlorinated polymers fade.

The solvatochromic parameters of the renewable diol derivatives are given in Table 13. The positioning of the diol derivatives on the Kamlet-Taft β vs π* plots has been examined (Figure 3). TMD and ETMD, as far as their polarity-basicity properties are concerned, are found to be close to such the aprotic solvents as diethyl ether, *n*-butyl acetate, eucalyptol and 1,1-diethoxymethane, and thus might be related to the group of solvents with the moderate basicity and the moderate polarity (Figure 3a). The higher boiling point and the lower volatility that may propose the lower evaporation losses and the lower flammability hazards are of potential benefit over Et_2O and DEM.

Table 12. The equilibrium solubilities of the organic polymers in the 1,2-PD and 2,3-BD derivatives.

Polymer	Equilibrium Concentration (25 °C), g L^{-1}			
	TMD	ETMD	IPP	SBB
Polystyrene	>300.0	>300.0	0.2	0.4
Polybutadiene	>300.0	>300.0	1.8	1.2
CPVC	8.9	11.3	1.9	1.7

ETMD is totally comparable with BuOAc in terms of volatility; the remarkable difference between these compounds is that the former (being the cyclic ketal) is stable in basic media, while the latter undergoes rapid saponification. Thus, the ketal solvents may be employed in those cases, when an organic solvent should be used in contact with the aqueous alkali solution. One should note that it was impossible for us to measure the E_T^N value for these compounds, since the Reichardt's dye solutions gave the inadequate wavenumbers, probably due to the presence of some minor impurities, which were not obliged to detecting by neither NMR nor GC. The determination of this abnormal behavior is of the further interest for the solvatochromic characterization of the cyclic ketals.

Table 13. The Kamlet-Taft solubility parameters for the renewable diol derivatives.

Solvent	λ_{max}, nm	E_T^N	λ_{max}, nm	π^*	λ_{max}, nm	β	λ_{max}, nm	α
IPP	572.2	0.595	304.8	0.56	367.6	0.77	572.2	0.81
SBB	626.2	0.462	303.3	0.49	364.9	0.77	626.2	0.58
TMD [1]	-	-	300.5	0.36	349.8	0.50	-	-
ETMD [1]	-	-	300.2	0.35	350.2	0.52	-	-

[1] The data on α and E_T^N is absent due to the abnormal behavior of the Reichardt's dye in the ketal samples upon the spectra acquisition.

(a)

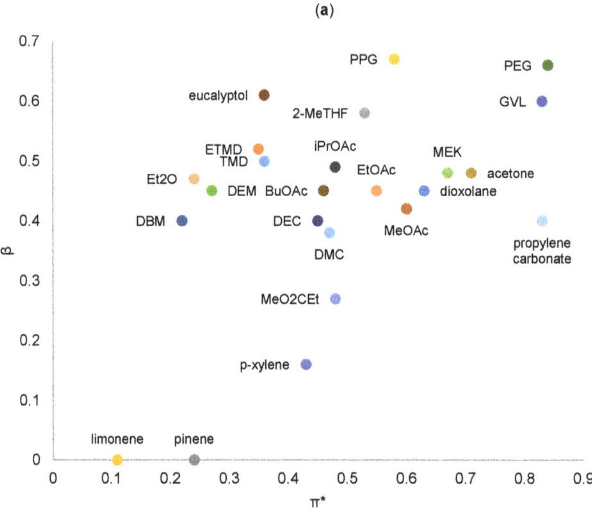

Figure 3. *Cont.*

(b)

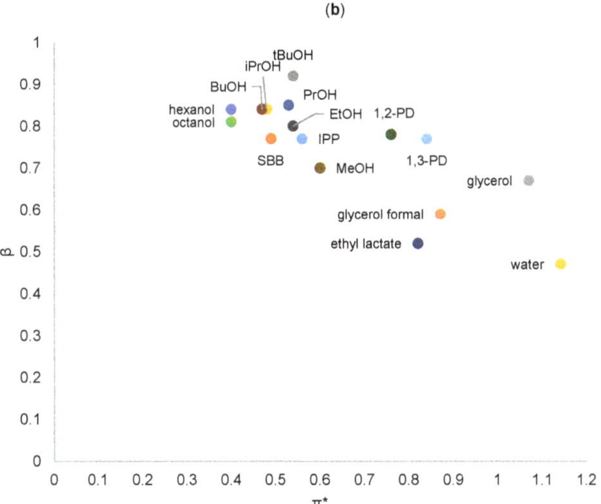

Figure 3. The positioning of the 1,2-PD and 2,3-BD derivatives on the β vs π* plots for aprotic (**a**) and protic (**b**) organic green solvents. The solvatochromic parameters for the rest of the compounds have been taken from the Ref. [60]. PEG—polyethylene glycol, GVL—γ-valerolactone, PPG—polypropylene glycol, DMC—dimethyl carbonate, DEC—diethyl carbonate, DEM—diethoxymethane, DBM—dibutoxymethane, 1,3-PD—1,3-propanediol, 2-MeTHF—2-methyltetrahydrofuran, MeO2CEt—methyl propionate.

SBB and IPP on the Kamlet-Taft plot for the protic solvents have been found among the aliphatic alcohols (Figure 3b). The boiling points of the glycol ethers are slightly lower than those of aliphatic alcohols with the same carbon atom content: the boiling points for IPP and 1-hexanol are 144 and 157 °C; 163 and 195 °C for SBB and 1-octanol, respectively. Simultaneously, in the aforementioned pairs the glycol ethers have higher miscibility with water, compared to the corresponding alcohols. While combining these two facets one can conclude that the closest alcohol analogues of IPP and SBB are 1-pentanol (bp = 138 °C, solubility in water 22 g L^{-1}) and 1-hexanol (bp = 157 °C, solubility in water 6 g L^{-1}), respectively. The main difference in case of IPP and pentanol is that the former is miscible with water in all the ratios. Hence, if the Ziegler process or the hydration of an oil-derived olefin is considered as the main source of fatty alcohols, IPP and SBB glycol ethers possibly have an advantage, since they are obtained from renewable resources, although 1-pentanol derived from levulinic acid might be bio-based as well.

The synthesis trees for the compounds tested in the given study (Figure 4) were made up based on the following assumptions: 2,3-butanediol is obtained by the microbial fermentation of carbohydrates with the subsequent one-step conversion to ETMD, according to the protocol described by Neish [23] and Harvey [22]; SBB is obtained by the one-step hydrogenolysis of ETMD. Thus, for these compounds the number of synthetic steps is two and three, respectively (Figure 4a). The synthetic trees for 1,2-PD derivatives have been made up starting from bioglycerol-derived propylene glycol.

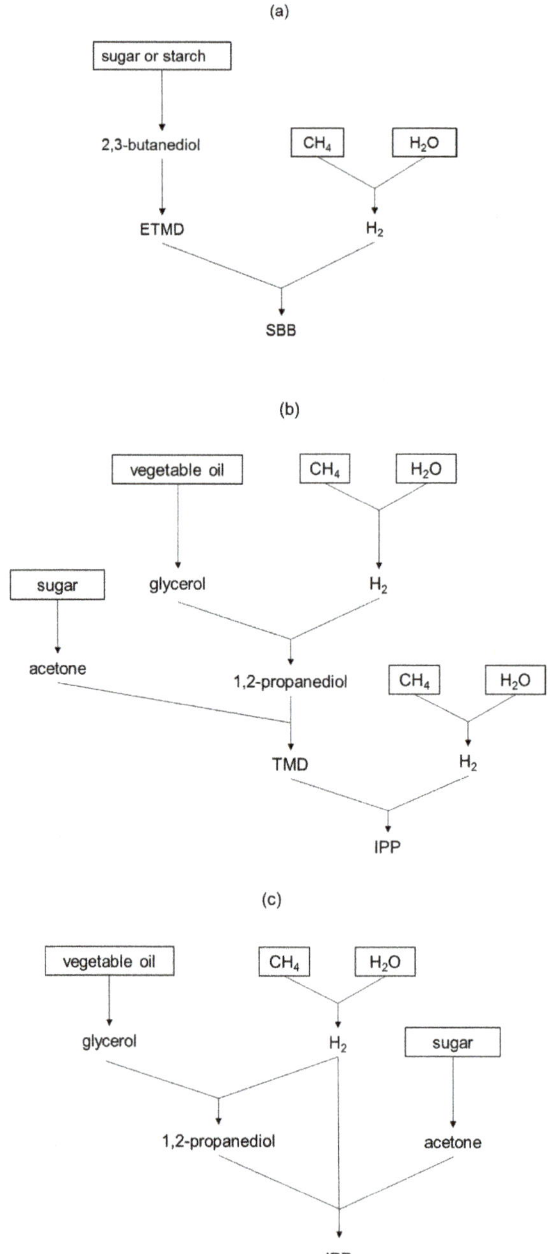

Figure 4. The synthesis trees for the renewable glycol derivatives: (**a**) ETMD and SBB, (**b**) IPP via the 1,2-PD ketalization and the TMD hydrogenolysis, (**c**) IPP via the reductive alkylation.

Although there are precedents for a one-step 1,2-PD synthesis by a retroaldol glucose conversion [61,62], currently only the glycerol hydrogenolysis process is employed on an industrial scale. Carbohydrate fermentation is supposed to be an appropriate source of the renewable acetone. The synthesis of TMD is implemented via the direct ketalization of 1,2-PD with acetone, since the feasibility of the one-step 1,2-PD to TMD conversion seems to be rather arguable in essence, representing here the dehydration of 1,2-PD to acetone. Thus, there are three synthetic steps in the TMD synthesis. For IPP obtained by the TMD hydrogenolysis it amounts to four (Figure 4b), but if the reductive alkylation route were chosen, IPP also would require only three steps to be recovered (Figure 4c).

In the final stage of our primary sustainability assessment for TMD, ETMD, IPP and SBB as organic solvents, the questions about the synthetic process formulated by Jessop [58] have been here addressed (Table 14). It is obvious that there are neither phosphorous- or nitrogen-containing wastes nor volatile heteroatomic compounds of nitrogen, sulfur and halogens in the production processes. The compounds under investigation might be considered to be fully (ETMD) or partially (TMD, IPP) renewable ones. Though providing the excellent yields and selectivity in rather mild conditions, the using of palladium catalysts for the cyclic ketal hydrogenolysis does not actually seem sustainable enough, thus employing a non-noble metal catalyst (e.g., nickel-based) for this reaction is of interest. Except the latter issue, the results of the primary sustainability estimate make it possible to suppose that the compounds investigated, particularly 2,3-butanediol-based, might be tentatively considered as potential green solvents.

Table 14. The considerations of the renewable diols derivatives production sustainability.

Parameter	TMD	ETMD	IPP	SBB
Number of synthetic steps	3	2	3	3
Phosphorous compounds in the synthesis tree		none		
Highly hazardous compounds		none		
Volatile compounds of N, S, Cl, Br, F		none		
Very hydrophobic compounds with logKow > 4.3		none		
Compound toxicity		no data		
Halogenated VOCs		none		
Elements at the risk of depletion in the synthesis	none		yes, palladium	
C_2–C_7 VOCs	yes, TMD	none	yes, IPP	none
Potentially renewable		yes		

As reported by Harvey et al., ETMD might have some potential as a gasoline component thanks to its appropriate volatility, relatively high calorific value and antiknock performance [22]. At the same time, there is no data on the influence of ETMD additives on the gasoline volatility properties and the antiknock performance. One can suppose that the hydrogenolysis of the cyclic ketals such as TMD and ETMD might give derivatives with increased antiknock performance, since the latter compounds bear an alcohol moiety. Alcohols are likely to be mostly efficient octane boosters when they are added to gasoline. Finally, the ketals possess relatively low stability in contact with acidic water [27,49], and despite the fact there is no data on the possible degradation of ketal-containing gasoline blends, this issue should not be disregarded. One should note here that the ethers which derive from ketal hydrogenolysis might have far higher hydrolytic stability than their precursors. Therefore, for characterizing the synthesized compounds in terms of the gasoline-blending properties the corresponding tests have been performed (Table 15). The extra purpose here was to estimate the relationships between the molecular structures of these oxygenates and their octane-enhancing efficiency.

As the tested oxygenates' boiling points are inside the gasoline boiling range (35–193 °C), their volatilities can be considered acceptable. The main concern here is that the "excessively heavy" oxygenates can decrease the overall gasoline volatility, what could be particularly relevant for high-boiling solketal [53], γ-valerolactone [63], methyl pentanoate and alkyl levulinates [64]. The results of the fractional composition, namely, the 5 and 10 vol. % recovery temperatures allow to make the "indirect" assertion on the gasoline cold start properties. As the differences in the $T_{5\%}$ and $T_{10\%}$ between the neat and additive-containing gasolines are within the method precision, one can conclude that none of the compounds tested affect the gasoline volatility.

Table 15. The gasoline blending properties of 1,2-PD and 2,3-BD ethers and ketals.

Property	Neat Gasoline	10% TMD	10% ETMD	10% IPP	10% SBB
d_{20}, kg L^{-1}	0.731	0.748	0.748	0.747	0.746
NHOC, MJ L^{-1}	32.31	31.64	31.91	31.714	31.92
RON	91.8	92.5	92.0	92.7	92.7
bRON	-	98.8	93.8	100.8	100.8
MON	84.5	85.2	85.2	85.9	85.8
bMON	-	91.5	91.5	97.5	98.5
blending AKI [1]		95.2	92.7	99.2	99.7
fractional composition, vol. %/°C					
ipb	35	36	40	36	35
5	45	49	50	46	44
10	52	57	57	53	50
20	61	68	73	68	63
30	72	80	86	80	80
40	85	91	103	94	94
50	104	107	121	114	114
60	124	120	135	127	135
70	142	136	146	139	149
80	160	157	160	150	161
90	186	192	187	175	182
fbp	193	194	190	190	189

[1] The deriving made from the calculated bON values.

The blending octane numbers and the calculated AKI for ETMD (Table 15) appeared to be slightly higher than the intrinsic ONs, reported previously by Harvey et al. [22]. Other tested oxygenates showed excellent antiknock performance enhancing the knock resistance more efficiently than ETMD: the AKI values for TMD, IPP and SBB amounted to 95.2, 99.2 and 99.7 points, respectively. The impacts on the RON are not very high in comparison with, e.g., ethanol and MTBE (blending RONs about 110–120 and 115–120, respectively), whereas the calculated values of bMON are comparable. Oxygenates providing a higher effect on MON are particularly preferable, when gasoline contains significant amounts of FCC gasoline, which normally has a high octane sensitivity. As expected, TMD turned out to have a higher octane number than ETMD: for the former, there is only one atom at the tertiary carbon per molecule, while for the latter there are two, being less resistant to oxidation.

It is of big concern that the molecular structure of TMD resembles the solketal molecule but without an OH-group. The methylation of the solketal OH-group was shown earlier to result in a dramatic ON decrease [53], and it was the OH-group (not the 2,2-dimethyl-1,3-dioxolane moiety) which was responsible for the good antiknock performance. In the present case, the octane rating of the 2,2-dimethyl-1,3-dioxolane derivative is obviously quite high. The understanding of the data on the solketal and its methylated derivative should be thus reconsidered: it seems that it is not only the positive influence of the OH-group in the solketal molecule, but the methoxy substituent is sure to affect the antiknock performance as well (Figure 5). An investigation on the relations between the molecular structure and the antiknock performance of 1,3-dioxolane homologues might be of further interest.

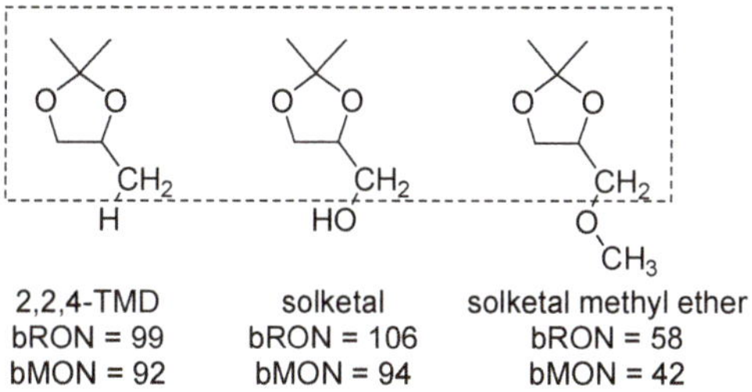

2,2,4-TMD   solketal   solketal methyl ether
bRON = 99   bRON = 106   bRON = 58
bMON = 92   bMON = 94   bMON = 42

Figure 5. The molecular structure and the octane numbers of the 1,3-dioxolane derivatives. The octane numbers of solketal and solketal methyl ether were taken from the ref. [53].

The mild hydrogenation of the cyclic ketals occurs with the release of the free alcohol group which might be considered as a "bearer" of the antiknock properties, as has been reported earlier [53,65]. There is little difference between the efficiency of the glycol ethers. SBB shows a slightly higher performance, although the higher impact from IPP having fewer C-H bonds at tertiary carbon atoms might be expected. However, "octane number–additive concentration" dependencies often demonstrate nonlinearity, that is why IPP and SBB might be considered to demonstrate a similar antiknock performance with the volume bMON/bRON about 101/98. The calculated molar bRON/bMON for the aforementioned glycol ethers are 101/98 and 100/97, respectively.

One should note that the volumetric calorific value of the renewable diols derivatives is quite high, compared to the alcohols with close antiknock performance. For example, 2-butanol with RON/MON = 105/93 has a density (at 20 °C) of 0.808 kg L^{-1} and a volumetric NHOC of 26.7 MJ L^{-1}. The weight calorific value of SBB (32.5 MJ kg^{-1}) might also be compared with that of 2-butanol and 2-methyl tetrahydrofuran (32.9 MJ kg^{-1}). The SBB volumetric NHOC amounts to 28.48 MJ L^{-1} with an antiknock performance similar to that of the aforementioned alcohol [64]. Upon adding 10 vol. % of 2-butanol and SBB, the volumetric NHOC values for the blends might amount to 26.70 and 28.48 MJ L^{-1} with AKI of the both blends of 89.2–89.3 points. Thus, the loss in the overall fuel volumetric calorific value upon the adding of the octane booster is remarkably lower for SBB. What is interesting, is that both the compounds could be obtained by bio-2,3-butanediol treatment involving the dehydration with the subsequent hydrogenation. On the Van Krevelen diagram the coordinates ([O:C; H:C]) of SBB and 2-BuOH are [2.25; 0.25] and [2.50; 0.25], respectively. Thus, although alcohol has a higher hydrogen content, the efficiency of SBB as the gasoline component is higher, so one can make the conclusion that the conversion of 2,3-BD to SBB represents a method to the renewable energy recovery which might be potentially efficient and sustainable.

3. Materials and Methods

3.1. Reagents

Acetone, MEK (analytical grade, Komponent-Reaktiv, Moscow, Russia), 1,2-propanediol (>99%, Carl Roth, Karlsruhe, Germany), racemic 2,3-butanediol (Alfa Aesar, Heysham, United Kingdom, 98%), LAH (95%, Sigma-Aldrich, St. Louis, MO, USA), AlCl$_3$ (Sigma-Aldrich, St. Louis, MO, USA, 99%), NaOH (pure, Komponent-Reaktiv, Moscow, Russia) diethyl ether (99.8%, Sigma-Aldrich, St. Louis, MO, USA) were used without the additional purification. The base gasoline without any oxygenates of additives was received from a Russian refinery. The properties of the base gasoline are given in Table S2 in the Supplementary Materials.

3.2. Preparative Syntheses

3.2.1. Cyclic Ketal Syntheses

To prepare cyclic ketals (TMD and ETMD), the diol was mixed with the ketone, taken in a 6-fold molar excess, in a round-bottom flask. A dash of *p*-toluenesulfonic acid (*p*-TSA) was introduced into the flask. The mixture was stirred at room temperature for 12 h, after which the catalyst was neutralized with an excess of sodium carbonate. The reaction mixture, which consisted of the unreacted ketone and diol, water and the ketal, was subjected to rectification to obtain the pure ketal. The purity and structure of the isolated ketal was checked by GC-FID ('Kristalluks-4000M' gas chromatograph, Meta-Khrom, Yoshkar-Ola, Russia) equipped with a 'Restek Rtx-Wax' 60 m/0.53 mm/1.00 μm column, Restek, Bellefonte, PA, USA, helium as the carrier gas, pure samples used as the internal standards for the quantification, 'Supelco SPB-1' 30 m/0.32 mm/0.25 μm column, Supelco, Bellefonte, PA, USA, for the RI determination), GC–MS ('Thermo Focus DSQ II', Thermo Fischer Scientific, Waltham, MA, USA, 'Varian VF-5ms' 30 m/0.25 mm/0.25 μm column, Varian Inc., Palo Alto, CA, USA, helium as the carrier gas, the EI ionization), the chemical analysis ('Thermo Flash 2000', Thermo Fischer Scientific, Waltham, MA, USA), and proton magnetic resonance ('Avance 400' Bruker, Billerica, MA, USA).

2,2,4-Trimethyl-1,3-dioxolane (TMD): bp 98–99 °C, yield 83.7 mol % (isolated, on a theoretical basis). Found (%): C, 62.11; H, 10.48. Calculated for $C_6H_{12}O_2$ (%): C, 62.04; H, 10.41. ^{1}H–NMR (400 MHz, CDCl$_3$): δ 1.18 (d, J_{HH} = 6.0 Hz, 3H), 1.35 (s, 3H), 1.44 (s, 3H), 3.34-3.38 (m, 1H), 3.95–3.99 (m, 1H), 4.12–4.19 (m, 1H). ^{13}C–NMR (100 MHz, CDCl$_3$): δ 18.4, 26.1, 26.4, 70.8, 72.1, 108.6. The mass spectrum, *m/z* (Irel, %): 115 (47), 72 (23), 59 (18), 43 (100). GC: non-polar column, temperature ramp, Van den Dool and Kratz RI = 721, (poly(dimethylsiloxane), 30 m/0.32 mm/0.25 μm, He, 45 to 365 °C, 25 °C/min). UV-Vis spectroscopy: no band with ε more than 100 has been found in the spectrum range of 300–630 nm.

2-Ethyl-2,4,5-trimethyl-1,3-dioxolane (ETMD): bp 135–142 °C, yield 81.6 mol % (isolated, on a theoretical basis). Found (%): C, 66.69; H, 11.08. Calculated for $C_8H_{16}O_2$ (%): C, 66.63; H, 11.18. The spectral data matched the literature data [66]. GC: non-polar column, the temperature ramp, Van den Dool and Kratz RI = 893, (poly(dimethylsiloxane), 30 m/0.32 mm/0.25 μm, He, 45 to 365 °C, 25 °C/min). UV-Vis spectroscopy: no band with ε more than 100 has been found in the spectrum range of 300–630 nm.

3.2.2. The Synthesis of Glycol Ethers by Cyclic Ketal Reduction

The reduction of cyclic ketals with LAH-AlCl$_3$ in ether was performed according to the protocol described by Legetter and Brown [29]. The products obtained after the ether distillation were double distilled before performing purity analysis.

Isopropoxypropanols (IPP, a mixture of 2-isopropoxy-1-propanol and 1-isopropoxy-2-propanol): bp 137–144 °C, yield 89.7 mol % (isolated, on a theoretical basis). Found (%): C, 60.87; H, 11.88. It has Calculated for $C_6H_{14}O_2$ (%): C, 60.98; H, 11.94. ^{1}H–NMR (400 MHz, CDCl$_3$): δ 1.09–1.11 (m, 6H), 1.13–1.19 (m, 3H), 2.89–3.01 (m, 1H), 3.42–3.64 (m, 2H), 3.89–3.96 (m, 1H). ^{13}C–NMR (100 MHz, CDCl$_3$): δ 19.0 (17.7), 21.7 (22.2), 66.5 (66.0), 71.4 (70.6), 73.7 (72.5). The MS data matched the NIST main library spectra. GC: non-polar column, the temperature ramp, Van den Dool and Kratz RI = 800, (poly(dimethylsiloxane), 30 m/0.32 mm/0.25 μm, He, 45 to 365 °C, 25 °C/min). UV-Vis spectroscopy: no band with ε more than 100 has been found in the spectrum range of 300–630 nm.

3-(sec-Butoxy)butan-2-ol (SBB, a mixture of the diastereomers): bp 162–163 °C, yield 92.0 mol % (isolated, on a theoretical basis). Found (%): C, 65.88; H, 12.38. Calculated for $C_8H_{18}O_2$ (%): C, 65.71; H, 12.41. ^{1}H–NMR (400 MHz, CDCl$_3$): δ 0.91 (t, J_{HH} = 7.4 Hz, 3H), 1.08–1.13 (m, 6H), 1.16 (d, J_{HH} = 6.4 Hz, 3H), 1.37–1.43 (m, 1H), 1.57–1.63 (m, 1H), 3.45–3.51 (m, 1H), 3.54–3.58 (m, 1H), 3.72–3.77 (m, 1H). ^{13}C–NMR (100 MHz, CDCl$_3$): δ 10.7, 16.5, 18.3, 19.4, 29.5, 70.2, 73.6, 76.6. Mass-spectrum, *m/z* (I_{rel}, %): 101 (56), 73 (24), 57 (100), 55 (32), 45 (74). GC: non-polar column, temperature ramp, Van den Dool and

Kratz RI = 984, (poly(dimethylsiloxane), 30 m/0.32 mm/0.25 μm, He, 45 to 365 °C, 25 °C/min). UV-Vis spectroscopy: no band with ε more than 100 has been found in the range of 300–630 nm.

3.3. Catalytic Experiments

The experiments of ketalization of the diols (1,2-PD and 2,3-BD) with ketones (acetone and MEK) under atmospheric pressure were carried out in a glass reactor (internal volume, 15 cm^3) equipped with a thermostating jacket and a reflux condenser. A magnetic stir bar and the previously prepared reactant mixture (7 mL) were placed in the reactor and the thermostat was turned on. After reaching the set temperature, a catalyst charge was introduced into the reactor. This moment was taken as the initial reaction time. Stirring was periodically terminated to take aliquots of a 0.05–0.1 mL volume, which were analyzed by GLC. The following sampling protocol was applied: for each reaction equilibrium state 5–7 samples were taken; each sample was analyzed from 3 to 5 times to obtain the appropriate standard deviation (it did not exceed 0.2% for the diol conversion values from 65 to 95%).

The experiments on the cyclic ketals hydrogenolysis and the reductive alkylation were carried out in a stainless steel autoclave (internal volume of 50 mL, designed and manufactured by TIPS RAS, Moscow, Russia), equipped with a manometer (for pressure control) and a thermocouple. The specified amounts of the cyclic ketal (TMD or ETMD in the ketal hydrogenolysis) or the mixture of the diol and the ketone (acetone+1,2-PD or MEK+2,3-BD in the reductive alkylation), the catalyst sample, and a magnetic stir bar were placed in the autoclave. The aluminosilica-based Al-HMS catalysts were used in the oxide form after the calcination. The closed autoclave was purged with argon, then filled with hydrogen (to an initial pressure of 40 bar), and then placed in a furnace. The stirring was turned on, and this point of the time was taken as the onset of the reaction. At the end of the experiment time, the autoclave was cooled with the cold water, and the products of the catalysis were decanted, centrifuged and subjected to analysing by GC-FID (the pure samples used as the internal standards for quantification) and GC–MS. The catalyst was washed five times with acetone (the suspension was centrifuged and acetone was decanted each time), and then dried in a flow of argon. The heterogeneous catalyst, thus prepared, was used for the TEM study.

In the test of the catalyst reusability, after each cycle the catalyst was recovered in the same manner as for the characterization. The equal feed-to-catalyst weight ratio was maintained throughout the tests.

3.4. Catalysts

The commercially available sulfonic ion-exchange resin Amberlyst 36 dry (a sample has been kindly provided for testing by Dow Europe GmbH, Moscow, Russia) was used in the ketalization experiments without any prior modification. The preparation method details and the properties of Pd/Al-HMS catalyst have been reported elsewhere [39].

The textural characteristics of the synthesized aluminosilicates and the commercial supports were determined on the ASAP 2020 instrument from the company Micromeritics (Norcross, GA, USA) with the use of the low-temperature adsorption of nitrogen. The parameters were calculated by the BET method using the instrument software. The structure and morphology of the supported catalyst samples were studied by the high-resolution transmission electron microscopy (TEM) on a JEM 2100 electron microscope (from JEOL, Tokyo, Japan) at the accelerating voltage of 200 kV. The acid properties of the supports were studied using the temperature-programmed desorption of ammonia (TPD of NH$_3$). The experiments were performed on an USGA-101 sorption analyzer from Unisit (Moscow, Russia). Before the adsorption of ammonia, the samples were in situ calcined in the flow of the dry air at 500 °C for 1 h and then cooled to 60 °C in the flow of nitrogen. The saturation was carried out in the flow of ammonia diluted with helium (1:1) at 60 °C for 30 min. The physically adsorbed ammonia was removed at 100 °C in the flow of helium for 1 h. The thermal desorption curves of ammonia were obtained at the temperature range from 100 °C to 800 °C in the flow of helium at the heating rate of 8 K/min.

3.5. Methods for Estimating Gasoline and Solvent Properties

3.5.1. Physico-Chemical Properties

The main methods employed in the study on the solvent and gasoline blending properties have been summarized in Table 16.

The gross calorific values of the compounds were measured using a C200 calorimeter (IKA, Königswinter, Germany) according to the DIN 51900. The net calorific values were calculated regarding the hydrogen weight contents determined by the CHNS analysis.

In the water solubility measurements, water aliquots (0.01 g each) were added to a 3 mL aliquot of the organic compound with the periodical shaking until the phase separation had been visually observed. The similar procedure was used to measure the solubility of the organic compounds in water. To estimate polymer solubility, the specified polymer mass was brought in contact with the solvent (2.0 mL) with stirring at the temperature of 50–60 °C for 8 h.

Table 16. The methods used for the diol derivatives characterization.

Property	Method	Instrument	Method Precision
Density at 20 °C	ASTM D4052	VIP-2MR vibration densitometer	± 0.0001 g/cm^3
Dynamic viscosity at 20 °C	ASTM D445	VPZh-2 glass viscosimeter	± 0.0001 mm^2*s^{-1}
Refractive index at 20 °C n_D^{20}	-	IRF-22 refractometer	± 0.0001
Melting point	DIN 51421	Kristall-20E	± 1 °C
Saturated vapor pressure	ASTM D6378	Reid bomb, thermostate	± 1.2 kPa
Fractional composition	ASTM D86	ARN-PKhP	± 1 °C
RON/MON	ASTM D2700/ASTM D 2699	UIT-85	± 0.1

An aliquot of the supernatant liquid was taken off and then evaporated in an argon flow (T = 150 °C, for 4 h) to give the residual mass of the polymer dissolved. Polystyrene ("402" trademark, Nizhnekamskneftekhim, Nizhnekamsk, Russia, M_w = 422280, M_n = 847060), polybutadiene ("CKDL" trademark, Nizhnekamskneftekhim, Nizhnekamsk, Russia, M_w = 680870, M_n = 1470180) and chlorinated poly(vinylchloride) ("PSH-LS" trademark chlorinated polyvinylchloride resin, Kaustik, Sterlitamak, Russia) were used as the polymer samples. logK$_{ow}$ had been estimated using the CLogP group contribution method.

The saturated vapor pressure (SVP)—temperature dependences were obtained by the vacuum rectification of the compounds under the different residual pressures using the laboratory rectification column. The Antoine equation coefficients were calculated by means of the mathematical regression analysis.

The TGA evaporation rate measurements have been conducted using TGA/DSC3+ (Mettler Toledo, Columbus, OH, USA), with a TGA/HT DSC HSS2 detector. In a typical run the temperature was risen from 25 °C to 500 °C (the temperature ramp, 5 °C/min) under ambient air conditions (the flow of 20 mL min^{-1}). The samples (20 μL volume) were put into the Pt crucibles (inner volume 70 μL) before the measurement. From the TGA curves, the temperatures corresponding to the 90% weight loss were determined. Butyl acetate (99.5%, Sigma Aldrich, St. Louis, MO, USA) was used as the reference substance.

3.5.2. Solvatochromic Parameters Estimation

The parameter E$_T$ was determined using 2,6-diphenyl-4-(2,4,6-triphenyl-1-pyridino)-phenolate (Reichardt betaine, 90%, Sigma-Aldrich, St. Louis, MO, USA) without the additional purification. 4-Nitroaniline (99%, Fluka, Buchs, Switzerland) and 4-nitroanisole (97%, Sigma Aldrich, St. Louis,

MO, USA) were used to determine the parameters β and π^*, respectively. All the ethers and ketals samples were stored over the 4A molecular sieves prior to the spectra acquisition in order to remove the trace water.

The concentration of the solvatochromic indicators was set at such the level that the absorption maximum was within the range of 0.2–1.0 absorbance units. The absorption spectra were recorded with the UV-3600 double-beam double-monochromator spectrophotometer (Shimadzu, Kyoto, Japan) equipped with a TCC-240A Peltier cell temperature control system. The temperature was maintained at a level of 25 ± 0.1 °C. The solutions were placed in quartz cells with optical path lengths of 10 or 1 mm (in the case of high background solvent absorption). The corresponding solvent without a solvatochromic indicator was used as the reference solution. The spectra were recorded within the range of 300–800 nm (depending on the indicator) at a spectral resolution of 1 nm. The distance between points in the spectrum was 0.1 nm. The obtained spectra were subjected to the Savitzky-Golay smoothening with the subsequent localization of the long-wave absorption band λ_{max} (the precision was not worse than 0.1 nm). The Dimroth-Reichardt energy (kcal/mol) was calculated using the following equation [67]:

$$E_T(30) = hcN_A\widetilde{v}_{maxB} = 28590/\lambda_{maxB} \tag{3}$$

where h is the Planck constant, c is the speed of light, N_A is the Avogadro number, v_{maxB} and λ_{maxB} are the wavenumber (m^{-1}) and wavelength (nm) at the Reichardt betaine absorption maximum, respectively. The normalized parameter E_T (30) in the solvent S was calculated by comparing the obtained value with the reference data for water and tetramethylsilane (TMS):

$$E_T^N = \frac{E_T(30)_S - E_T(30)_{TMS}}{E_T(30)_{H_2O} - E_T(30)_{TMS}} = \frac{E_T(30)_S - 30.7}{32.4} \tag{4}$$

The Kamlet-Taft parameters π^* and β were calculated using the literature data on the corresponding coefficients (b, s) for the location of the absorption bands of the studied solvatochromic indicators according to the equations [67]:

$$\pi^* = 0.427(34.12 - \widetilde{v}_{maxANS}) \tag{5}$$

$$\beta = \frac{31.10 - 3.14\pi^* - \widetilde{v}_{maxANI}}{2.79} \tag{6}$$

where $v_{max\,ANI}$ and $v_{max\,ANS}$ are the absorption maximum wavenumbers for 4-nitroaniline and 4-nitroanisole, respectively (in kK, 1 kK = 1000 cm^{-1}). The parameter α was determined from the data on π^* and the following equation:

$$\alpha = 0.186(10.91 - \widetilde{v}_{maxB}) - 0.72\pi^* \tag{7}$$

4. Conclusions

Synthetic strategies for the cyclic ketal and the glycol ether derivatives of the renewable diols 1,2-propylene glycol and 2,3-butane diol have been proposed. The ketalization reaction has been subjected to a thermodynamic analysis. The homological reactions of the cyclic ketal formation from the diol and the ketone has been shown to comply with the isoequilibrium relationship. The $\Delta H^0_r/\Delta S^0_r$ (kJ mol^{-1}; J mol^{-1} K^{-1}) values for the reactions (1,2-propylene glycol + acetone → 2,2,4-trimethyl-1,3-dioxolane-+ water and 2,3-butanediol + 2-butanone → 2-ethyl-2,4,5-trimethyl-1,3-dioxolane) have been found to be −16.6±1.0/−56.8±3.3 and −15.1±0.9/−51.1±3.1, respectively.

The hydrogenolysis of the two cyclic ketals—TMD and ETMD—has been studied with the use of heterogeneous bifunctional catalysts obtained by doping Al-HMS mesoporous aluminosilicas with palladium metal. The catalysts have been successfully used for the synthesis of the glycol ethers via two different routes (the cyclic ketal hydrogenolysis and the reductive alkylation of the diol with the ketone): in both the cases the target products have been obtained with yields up to 91–93 mol %;

the catalysts have been found to be reusable at least 5 times with a moderate decrease in activity (the yield of the target product on the 5th cycle had been registered as 84% of the first cycle yield).

The renewable diol derivatives (the two ketals—TMD and ETMD—and the two corresponding ethers—IPP and SBB) have been characterized by their solvent-relevant properties, including density, viscosity, boiling points, melting points, refraction coefficient and heat of combustion. The volatility properties have been investigated in order to obtain the relative evaporation rates (RER), as well as the Antoine equation coefficients.

TMD shows the possibility to evaporate considerably faster (RER = 1.30) than BuOAc (the reference substance), while ETMD, IPP and SBB may evaporate slightly slower (RERs 0.91, 0.84 and 0.73, respectively).

A solvent properties study has been conducted for TMD, ETMD, IPP and SBB, including their miscibility with organic solvents and water, their ability to dissolve polymers (polystyrene, polybutadiene and CPVC) and Kamlet-Taft solvatochromic parameters determinations. In the polymer dissolution tests, the cyclic ketals have been found to be remarkable solvents for polystyrene and polybutadiene: more than 300 g of the polymer per liter of the solvent could be dissolved. After positioning on β vs π^* plots the conclusion has been made that the cyclic ketals (TMD and ETMD) appear to have close solvent properties to BuOAc, diethyl ether, eucalyptol and 1,1-diethoxy methane. The hydrolytic stability in the aqueous alkali media might be a possible advantage over BuOAc, while in the comparison with diethyl ether the ketals demonstrate considerably lower flammability hazards and potential evaporation losses for their lower volatility. Glycol ethers, IPP and SBB have similar solvent properties as C_5-C_8 fatty alcohols, what permits the alcohols (produced from ethylene via the Ziegler process) to be a potentially renewable alternative.

From the synthesis trees designed for the four compounds, the number of synthetic steps for TMD, ETMD, IPP and SBB has been found to be three, two, three and three, respectively. The primary sustainability estimate made for the synthetic protocols has shown positive results for most of the criteria. The problem to be solved here is the necessity of using palladium (an element at risk of the depletion) for the glycol ether synthesis by ketal hydrogenolysis.

We have also estimated the gasoline-blending properties of the renewable ketal glycol ethers. The blending research/motor octane numbers for TMD, ETMD, IPP and SBB have been found equal to 98.8/91.5, 93.8/91.5, 100.1/97.5 and 100.8/98.5, respectively. Thus, after subjecting ETMD to a mild and highly selective hydrogenolysis it has yielded an oxygenate with a substantially higher performance both in terms of its antiknock properties and hydrolytic stability.

SBB, with the highest boiling point (163–164 °C) among the oxygenates tested has not been found to affect gasoline volatility. The comparison between two derivatives of 2,3-BD (2-butanol and SBB) has shown that although the antiknock properties of the two compounds appear to be totally comparable, SBB has a particular advantage as far as the volumetric calorific value is concerned. After adding 10 vol. % of 2-butanol and SBB, the fuel AKI has increased from 88.2 to 89.2–89.3 points, while the NHOC for the blends have been 26.70 and 28.48 $MJ\cdot L^{-1}$ (6–7% more for the last mentioned one), respectively.

From the results obtained in the given investigation, a number of prospective future research directions may be outlined. The cyclic ketals and the corresponding glycol ethers (besides IPP being essentially an industrially available product) are reasonably estimated to be good thinners for dyes and the coating formulations. There is an example of the successful application of solketal (2,2-dimethyl-4-hydroxymethyl-1,3-dioxolane), a bio-based cyclic ketal being currently produced and utilized on the industrial scale.

The lower polarity and hydrophobicity of the diol ketals compared to solketal may help some of these compounds find applications in adjacent niches. For example, TMD and ETMD could be tested as possible replacements for conventional solvents such as THF and xylenes. One should mention SBB as the renewable isomer of petroleum-derived ethylene glycol monohexyl ether. Of course, the thorough toxicity and the environmental impact estimate are of particular interest.

Both the ketal hydrogenolysis and reductive alkylation approaches seem rather advantageous in terms of reaction selectivity and atom efficiency, except for the need of using a Pd-containing catalyst. It is thus of interest to substitute palladium with more sustainable alternatives, for example, nickel. These catalysts may be useful for homologous reactions, e.g., the reductive alkylation of furfural with lower alcohols.

Supplementary Materials: The following are available online, Figure S1: Evaporation rates vs saturated vapor pressure (under 50 °C) plot, Figure S2: Evaporation rates vs boiling point plot; Table S1: Reduced pressure boiling points for TMD, ETMD, IPP and SBB; Table S2: Main physico-chemical properties of the base gasoline.

Author Contributions: Conceptualization, V.S. and A.M.; methodology, V.S., A.L. and D.K.; investigation, D.N., A.G., D.Z., M.K., A.L. and M.B.; writing—original draft preparation, V.S.; writing—review and editing, V.S. and A.M.; supervision, A.M., M.B. and D.K.; project administration, A.M.; funding acquisition, V.S. All authors have read and agreed to the published version of the manuscript.

Funding: The reported study has been funded by Russian Science Foundation (RSF) according to the research project No. 18-73-00187.

Acknowledgments: The work has been carried out using the instrumentation of the Core Facility Center "Arktika" of the Northern (Arctic) Federal University named after M.V. Lomonosov. The authors here express their gratitude to Roman S. Borisov for the help in conducting the GC-MS analysis and the particular gratitude to Vasily A. Demidov (Dow Europe GmbH, Moscow, Russia) for supplying us with the Amberlyst 36 ion-exchange resin.

Conflicts of Interest: The authors declare no conflict of interest.

Abbreviations

AKI: antiknock index; 2,3-BD, 2,3-butanediol; CPVC, chlorinated poly(vinyl)chloride; ETMD, 2-ethyl-2,4,5-trimethyl-1,3-dioxolane; 1,2-PD, 1,2-propanediol; IPP, isopropoxypropanol (an isomeric mixture of 2-isopropoxy-1-propanol and 1-isopropoxy-2-propanol); MEK, methyl ethyl ketone; MON, motor octane number; bMON, volumetric blending motor octane number; NHOC, net heat of combustion; RON, research octane number; bRON, volumetric blending research octane number; SBB, 3-(sec-butoxy)butan-2-ol (the mixture of diastereomers); SD, standard deviation; SVP, saturated vapor pressure; RER, relative (to *n*-butyl acetate) evaporation rate; TMD, 2,2,4-trimethyl-1,3-dioxolane.

References

1. Nachtergaele, P.; De Meester, S.; Dewulf, J. Environmental sustainability assessment of renewables-based propylene glycol at full industrial scale production. *J. Chem. Technol. Biotechnol.* **2019**, *94*, 1808–1815. [CrossRef]

2. Gonzalez-Garay, A.; Gonzalez-Miquel, M.; Guillen-Gosalbez, G. High-value propylene glycol from low-value biodiesel glycerol: A techno-economic and environmental assessment under uncertainty. *Acs Sustain. Chem. Eng.* **2017**, *5*, 5723–5732. [CrossRef]

3. Bennett, G.N.; San, K.Y. Microbial formation, biotechnological production and applications of 1,2-propanediol. *Appl. Microbiol. Biotechnol.* **2001**, *55*, 1–9. [CrossRef]

4. Ji, X.J.; Huang, H.; Ouyang, P.K. Microbial 2,3-butanediol production: A state-of-the-art review. *Biotechnol. Adv.* **2011**, *29*, 351–364. [CrossRef]

5. Białkowska, A.M. Strategies for efficient and economical 2,3-butanediol production: New trends in this field. *World J. Microbiol. Biotechnol.* **2016**, *32*. [CrossRef]

6. Koutinas, A.A.; Yepez, B.; Kopsahelis, N.; Freire, D.M.G.; de Castro, A.M.; Papanikolaou, S.; Kookos, I.K. Techno-economic evaluation of a complete bioprocess for 2,3-butanediol production from renewable resources. *Bioresour. Technol.* **2016**, *204*, 55–64. [CrossRef]

7. Zeng, A.P.; Sabra, W. Microbial production of diols as platform chemicals: Recent progresses. *Curr. Opin. Biotechnol.* **2011**, *22*, 749–757. [CrossRef]

8. Oliver, J.W.K.; Machado, I.M.P.; Yoneda, H.; Atsumi, S. Cyanobacterial conversion of carbon dioxide to 2,3-butanediol. *Proc. Natl. Acad. Sci. USA* **2013**, *110*, 1249–1254. [CrossRef]

9. Nikitina, M.A.; Sushkevich, V.L.; Ivanova, I.I. Dehydration of 2,3-butanediol over zeolite catalysts. *Pet. Chem.* **2016**, *56*, 230–236. [CrossRef]

10. Nikitina, M.A.; Ivanova, I.I. Conversion of 2,3-Butanediol over phosphate catalysts. *Chem. Cat. Chem.* **2016**, *8*, 1346–1353. [CrossRef]

11. Zhao, J.; Yu, D.; Zhang, W.; Hu, Y.; Jiang, T.; Fu, J.; Huang, H. Catalytic dehydration of 2,3-butanediol over P/HZSM-5: Effect of catalyst, reaction temperature and reactant configuration on rearrangement products. *Rsc Adv.* **2016**, *6*, 16988–16995. [CrossRef]

12. Zhang, W.; Yu, D.; Ji, X.; Huang, H. Efficient dehydration of bio-based 2,3-butanediol to butanone over boric acid modified HZSM-5 zeolites. *Green Chem.* **2012**, *14*, 3441–3450. [CrossRef]

13. Duan, H.; Yamada, Y.; Sato, S. Vapor-phase catalytic dehydration of 2,3-butanediol into 3-buten-2-ol over Sc$_2$O$_3$. *Chem. Lett.* **2014**, *43*, 1773–1775. [CrossRef]

14. Duan, H.; Yamada, Y.; Kubo, S.; Sato, S. Vapor-phase catalytic dehydration of 2,3-butanediol to 3-buten-2-ol over ZrO$_2$ modified with alkaline earth metal oxides. *Appl. Catal. A Gen.* **2017**, *530*, 66–74. [CrossRef]

15. Duan, H.; Sun, D.; Yamada, Y.; Sato, S. Dehydration of 2,3-butanediol into 3-buten-2-ol catalyzed by ZrO$_2$. *Catal. Commun.* **2014**, *48*, 1–4. [CrossRef]

16. Zheng, Q.; Wales, M.D.; Heidlage, M.G.; Rezac, M.; Wang, H.; Bossmann, S.H.; Hohn, K.L. Conversion of 2,3-butanediol to butenes over bifunctional catalysts in a single reactor. *J. Catal.* **2015**, *330*, 222–237. [CrossRef]

17. Zheng, Q.; Grossardt, J.; Almkhelfe, H.; Xu, J.; Grady, B.P.; Douglas, J.T.; Amama, P.B.; Hohn, K.L. Study of mesoporous catalysts for conversion of 2,3-butanediol to butenes. *J. Catal.* **2017**, *354*, 182–196. [CrossRef]

18. Kwok, K.M.; Choong, C.K.S.; Ong, D.S.W.; Ng, J.C.Q.; Gwie, C.G.; Chen, L.; Borgna, A. Hydrogen-Free Gas-Phase Deoxydehydration of 2,3-Butanediol to Butene on Silica-Supported Vanadium Catalysts. *ChemCatChem* **2017**, *9*, 2443–2447. [CrossRef]

19. Liu, X.; Fabos, V.; Taylor, S.; Knight, D.W.; Whiston, K.; Hutchings, G.J. One-step production of 1,3-butadiene from 2,3-butanediol dehydration. *Chem. A Eur. J.* **2016**, *22*, 12290–12294. [CrossRef]

20. Song, D. Kinetic model development for dehydration of 2,3-butanediol to 1,3-butadiene and methyl ethyl ketone over an amorphous calcium phosphate catalyst. *Ind. Eng. Chem. Res.* **2016**, *55*, 11664–11671. [CrossRef]

21. Kim, W.; Shin, W.; Lee, K.J.; Song, H.; Kim, H.S.; Seung, D.; Filimonov, I.N. 2,3-Butanediol dehydration catalyzed by silica-supported sodium phosphates. *Appl. Catal. A Gen.* **2016**, *511*, 156–167. [CrossRef]

22. Harvey, B.G.; Merriman, W.W.; Quintana, R.L. Renewable gasoline, solvents, and fuel additives from 2,3-butanediol. *ChemSusChem* **2016**, *9*, 1–7. [CrossRef] [PubMed]

23. Neish, A.C.; Haskell, V.C.; Macdonald, F.J. Production and properties of 2,3-butanediol. VI. Dehydration by sulphuric acid. *Can. J. Res.* **1945**, *23*, 281–289. [CrossRef]

24. Staples, O.; Moore, C.M.; Leal, J.H.; Semelsberger, T.A.; McEnally, C.S.; Pfefferle, L.D.; Sutton, A.D. A simple, solvent free method for transforming bio-derived aldehydes into cyclic acetals for renewable diesel fuels. *Sustain. Energy Fuels* **2018**, *2*, 2742–2746. [CrossRef]

25. Kapkowski, M.; Popiel, J.; Siudyga, T.; Dzida, M.; Zorębski, E.; Musiał, M.; Sitko, R.; Szade, J.; Balin, K.; Klimontko, J.; et al. Mono- and bimetallic nano-Re systems doped Os, Mo, Ru, Ir as nanocatalytic platforms for the acetalization of polyalcohols into cyclic acetals and their applications as fuel additives. *Appl. Catal. B Environ.* **2018**, *239*, 154–167. [CrossRef]

26. Samoilov, V.O.; Ni, D.S.; Dmitriev, G.S.; Zanaveskin, L.N.; Maximov, A.L. The joint synthesis of 1,2-propylene glycol and isopropyl alcohol by the copper-catalyzed hydrogenolysis of solketal. *ACS Sustain. Chem. Eng.* **2019**, *7*, 9330–9341. [CrossRef]

27. Moity, L.; Benazzouz, A.; Molinier, V.; Nardello-Rataj, V.; Elmkaddem, M.K.; de Caro, P.; Thiébaud-Roux, S.; Gerbaud, V.; Marion, P.; Aubry, J.-M. Glycerol acetals and ketals as bio-based solvents: Positioning in Hansen and COSMO-RS spaces, volatility and stability towards hydrolysis and autoxidation. *Green Chem.* **2015**, *17*, 1779–1792. [CrossRef]

28. Leggetter, B.E.; Diner, U.E.; Brown, R.K. The relative ease of reductive cleavage of 1,3-dioxolanes and 1,3-dioxanes in ether solution by LiAlH$_4$–AlCl$_3$. *Can. J. Chem.* **1964**, *42*, 2113–2118. [CrossRef]

29. Leggetter, B.E.; Brown, R.K. The influence of substituents on the ease and direction of ring opening in the LiAlH$_4$ –AlCl$_3$ reductive cleavage of substituted 1,3-dioxolanes. *Can. J. Chem.* **1964**, *42*, 990–1004. [CrossRef]

30. Eliel, E.L.; Badding, V.G.; Rerick, M.N. Reduction with metal hydrides. 12. Reduction of acetals and ketals with lithium aluminum hydride-aluminum chloride. *J. Am. Chem. Soc.* **1962**, *84*, 2371. [CrossRef]

31. Bhattacharjee, S.S.; Gorin, P.A.J. Hydrogenolysis of carbohydrate acetals, ketals, and cyclic orthoesters with lithium aluminium hydride–aluminium trichloride. *Can. J. Chem.* **1969**, *47*, 1195–1206. [CrossRef]

32. Abdun-Nur, A.R.; Issidorides, C.H. Pentaerythritol derivatives. V.1 Preparation of diethers of pentaerythritol by reduction of acetals and ketals. *J. Org. Chem.* **1962**, *27*, 67–70. [CrossRef]

33. Purushothama Chary, K.; Santosh Laxmi, Y.R.; Iyengar, D.S. Reductive cleavage of acetals/ketals with ZrCl₄/NaBH₄. *Synth. Commun.* **1999**, *29*, 1257–1261. [CrossRef]

34. Nutaitis, C.F.; Gribble, G.W. Reactions of sodium borohydride in acidic media. XIV. Reductive cleavage of cyclic acetals and ketals to hydroxyalkyl ethers. *Org. Prep. Proced. Int.* **1985**, *17*, 11–16. [CrossRef]

35. Shi, Y.; Dayoub, W.; Chen, G.R.; Lemaire, M. TMDS-Pd/C: A convenient system for the reduction of acetals to ethers. *Tetrahedron Lett.* **2011**, *52*, 1281–1283. [CrossRef]

36. Zhang, Y.-J.; Dayoub, W.; Chen, G.-R.; Lemaire, M. Environmentally benign metal triflate-catalyzed reductive cleavage of the C–O bond of acetals to ethers. *Green Chem.* **2011**, *13*, 2737. [CrossRef]

37. Fleming, B.I.; Bolker, H.I. The reduction of acetals with cobalt carbonyl catalysts. *Can. J. Chem.* **1976**, *54*, 685–694. [CrossRef]

38. Shi, Y.; Dayoub, W.; Favre-Réguillon, A.; Chen, G.R.; Lemaire, M. Straightforward selective synthesis of linear 1-*O*-alkyl glycerol and di-glycerol monoethers. *Tetrahedron Lett.* **2009**, *50*, 6891–6893. [CrossRef]

39. Samoilov, V.; Onishchenko, M.; Ramazanov, D.; Maximov, A. Glycerol isopropyl ethers: Direct synthesis from alcohols and synthesis by the reduction of solketal. *ChemCatChem* **2017**, *9*, 2839–2849. [CrossRef]

40. Shi, Y.; Dayoub, W.; Chen, G.-R.; Lemaire, M. Selective synthesis of 1-*O*-alkyl glycerol and diglycerol ethers by reductive alkylation of alcohols. *Green Chem.* **2010**, *12*, 2189. [CrossRef]

41. Sutter, M.; Da Silva, E.; Duguet, N.; Raoul, Y.; Métay, E.; Lemaire, M. Glycerol ether synthesis: A bench test for green chemistry concepts and technologies. *Chem. Rev.* **2015**, *115*, 8609–8651. [CrossRef]

42. Bethmont, V.; Montassier, C.; Marecot, P. Ether synthesis from alcohol and aldehyde in the presence of hydrogen and palladium deposited on charcoal. *J. Mol. Catal. A Chem.* **2000**, *152*, 133–140. [CrossRef]

43. Moity, L.; Shi, Y.; Molinier, V.; Dayoub, W.; Lemaire, M.; Aubry, J.M. Hydrotropic properties of alkyl and aryl glycerol monoethers. *J. Phys. Chem. B* **2013**, *117*, 9262–9272. [CrossRef] [PubMed]

44. Queste, S.; Bauduin, P.; Touraud, D.; Kunz, W.; Aubry, J.-M. Short chain glycerol 1-monoethers—A new class of green solvo-surfactants. *Green Chem.* **2006**, *8*, 822–830. [CrossRef]

45. Cornejo, A.; Barrio, I.; Campoy, M.; Lazaro, J.; Navarrete, B. Oxygenated fuel additives from glycerol valorization. Main production pathways and effects on fuel properties and engine performance: A critical review. *Renew. Sustain. Energy Rev.* **2017**, *79*, 1400–1413. [CrossRef]

46. Leal-Duaso, A.; Pérez, P.; Mayoral, J.A.; Garciá, J.I.; Pires, E. Glycerol-derived solvents: Synthesis and properties of symmetric glyceryl diethers. *ACS Sustain. Chem. Eng.* **2019**, *7*, 13004–13014. [CrossRef]

47. Anteunis, M.; Rommelaere, Y. NMR Experiments on acetals. XXIX. The ease of acetonide formation of some glycols. *Bull. Des. Sociétés Chim. Belg.* **1970**, *79*, 523–530. [CrossRef]

48. Lorette, N.B.; Howard, W.L.; Brown, J.H. Preparations of ketone acetals from linear ketones and alcohols. *J. Org. Chem.* **1959**, *24*, 1731–1733. [CrossRef]

49. Ozorio, L.P.; Pianzolli, R.; Mota, M.B.S.; Mota, C.J.A. Reactivity of glycerol/acetone ketal (solketal) and glycerol/formaldehyde acetals toward acid-catalyzed hydrolysis. *J. Braz. Chem. Soc.* **2012**, *23*, 931–937. [CrossRef]

50. Da Silva, C.X.A.; Gonçalves, V.L.C.; Mota, C.J.A. Water-tolerant zeolite catalyst for the acetalisation of glycerol. *Green Chem.* **2009**, *11*, 38. [CrossRef]

51. Samoilov, V.O.; Ni, D.S.; Maximov, A.L. Transacetalization of solketal: A greener route to bioglycerol-based speciality chemicals. *Chem. Select.* **2018**, *3*, 9759–9766. [CrossRef]

52. Nanda, M.R.; Yuan, Z.; Qin, W.; Ghaziaskar, H.S.; Poirier, M.A.; Xu, C.C. Thermodynamic and kinetic studies of a catalytic process to convert glycerol into solketal as an oxygenated fuel additive. *Fuel* **2014**, *117*, 470–477. [CrossRef]

53. Samoilov, V.O.; Maximov, A.L.; Stolonogova, T.I.; Chernysheva, E.A.; Kapustin, V.M.; Karpunina, A.O. Glycerol to renewable fuel oxygenates. Part I: Comparison between solketal and its methyl ether. *Fuel* **2019**, *249*, 486–495. [CrossRef]

54. Dmitriev, G.S.; Terekhov, A.V.; Zanaveskin, L.N.; Khadzhiev, S.N.; Zanaveskin, K.L.; Maksimov, A.L. Choice of a catalyst and technological scheme for synthesis of solketal. *Russ. J. Appl. Chem.* **2016**, *89*, 1619–1624. [CrossRef]

55. Howard, W.L.; Brown, J.H. Hydrogenolysis of ketals. *J. Org. Chem.* **1961**, *26*, 1026–1028. [CrossRef]

56. García, J.I.; García-Marín, H.; Pires, E. Glycerol based solvents: Synthesis, properties and applications. *Green Chem.* **2010**, *12*, 426–434. [CrossRef]

57. Moity, L.; Molinier, V.; Benazzouz, A.; Joossen, B.; Gerbaud, V.; Aubry, J.M. A "top-down" in silico approach for designing ad hoc bio-based solvents: Application to glycerol-derived solvents of nitrocellulose. *Green Chem.* **2016**, *18*, 3239–3249. [CrossRef]

58. Jessop, P.G. Searching for green solvents. *Green Chem.* **2011**, *13*, 1391–1398. [CrossRef]

59. Queste, S.; Michina, Y.; Dewilde, A.; Neueder, R.; Kunz, W.; Aubry, J.-M. Thermophysical and bionotox properties of solvo-surfactants based on ethylene oxide, propylene oxide and glycerol. *Green Chem.* **2007**, *9*, 491. [CrossRef]

60. Jessop, P.G.; Jessop, D.A.; Fu, D.; Phan, L. Solvatochromic parameters for solvents of interest in green chemistry. *Green Chem.* **2012**, *14*, 1245–1259. [CrossRef]

61. Hirano, Y.; Sagata, K.; Kita, Y. Selective transformation of glucose into propylene glycol on Ru/C catalysts combined with ZnO under low hydrogen pressures. *Appl. Catal. A Gen.* **2015**, *502*, 1–7. [CrossRef]

62. Tan, Z.; Shi, L.; Zan, Y.; Miao, G.; Li, S.; Kong, L.; Li, S.; Sun, Y. Crucial role of support in glucose selective conversion into 1,2-propanediol and ethylene glycol over Ni-based catalysts: A combined experimental and computational study. *Appl. Catal. A Gen.* **2018**, *560*, 28–36. [CrossRef]

63. Horváth, I.T.; Mehdi, H.; Fábos, V.; Boda, L.; Mika, L.T. γ-Valerolactone-a sustainable liquid for energy and carbon-based chemicals. *Green Chem.* **2008**, *10*, 238–242. [CrossRef]

64. Christensen, E.; Yanowitz, J.; Ratcliff, M.; McCormick, R.L. Renewable oxygenate blending effects on gasoline properties. *Energy Fuels* **2011**, *25*, 4723–4733. [CrossRef]

65. Boot, M.D.; Tian, M.; Hensen, E.J.M.; Mani Sarathy, S. Impact of fuel molecular structure on auto-ignition behavior—Design rules for future high performance gasolines. *Prog. Energy Combust. Sci.* **2017**, *60*, 1–25. [CrossRef]

66. Khusnutdinov, R.I.; Shchadneva, N.A.; Burangulova, R.Y.; Muslimov, Z.S.; Dzhemilev, U.M. Oxidation of monohydric and dihydric alcohols with CCl_4 catalyzed by molybdenium compounds. *Russ. J. Org. Chem.* **2006**, *42*, 1615–1621. [CrossRef]

67. Ladesov, A.V.; Kosyakov, D.S.; Bogolitsyn, K.G.; Gorbova, N.S. Solvatochromic polarity parameters for binary mixtures of 1-butyl-3-methylimidazolium acetate with water, methanol, and dimethylsulfoxide. *Russ. J. Phys. Chem. A* **2015**, *89*, 1814–1820. [CrossRef]

Sample Availability: Samples of the compounds are available from the authors.

 *molecules* 

Article

The CO$_2$ Absorption in Flue Gas Using Mixed Ionic Liquids

Guoqing Wu, Ying Liu * , Guangliang Liu and Xiaoying Pang

College of Chemistry and Chemical Engineering, Inner Mongolia University, Hohhot 010021, China; 15771380268@163.com (G.W.); 111975317@imu.edu.cn (G.L.); Pmmwhydedabai@163.com (X.P.)
* Correspondence: celiuy@imu.edu.cn; Tel.: +86-471-4992981

Academic Editors: Monica Nardi, Antonio Procopio and Maria Luisa Di Gioia
Received: 22 January 2020; Accepted: 22 February 2020; Published: 25 February 2020

Abstract: Because of the appealing properties, ionic liquids (ILs) are believed to be promising alternatives for the CO$_2$ absorption in the flue gas. Several ILs, such as [NH$_2$emim][BF$_4$], [C$_4$mim][OAc], and [NH$_2$emim[OAc], have been used to capture CO$_2$ of the simulated flue gas in this work. The structural changes of the ILs before and after absorption were also investigated by quantum chemical methods, FTIR, and NMR technologies. However, the experimental results and theoretical calculation showed that the flue gas component SO$_2$ would significantly weaken the CO$_2$ absorption performance of the ILs. SO$_2$ was more likely to react with the active sites of the ILs than CO$_2$. To improve the absorption capacity, the ionic liquid (IL) mixture [C$_4$mim][OAc]/[NH$_2$emim][BF$_4$] were employed for the CO$_2$ absorption of the flue gas. It is found that the CO$_2$ absorption capacity would be increased by about 25%, even in the presence of SO$_2$. The calculation results suggested that CO$_2$ could not compete with SO$_2$ for reacting with the IL during the absorption process. Nevertheless, SO$_2$ might be first captured by the [NH$_2$emim][BF$_4$] of the IL mixture, and then the [C$_4$mim][OAc] ionic liquid could absorb more CO$_2$ without the interference of SO$_2$.

Keywords: flue gas; carbon dioxide; absorption; ionic liquids

---

1. Introduction

The CO$_2$ absorption of flue gas is an important process for reducing greenhouse gas [1]. To date, most flue absorptions are performed by using amine solvents. However, conventional absorption methods usually have some disadvantages: High equipment corrosion rate, high absorbent make-up rate due to the amine degradation by SO$_2$, NO$_2$, and O$_2$ in the flue gas, and high energy consumption during the regeneration process [2]. In the last decades, ionic liquids (ILs) have been used in many fields. Multifunctional ionic liquids are easily prepared, and the vapor pressure of ionic liquids can be neglected [3]. The other attractive properties of ILs include: High thermal stability, large electrochemical window, and high dissolve ability of compounds [4]. Blanchard et al. [5] have reported that certain ILs can considerably dissolve CO$_2$ gas. Since then, ILs for CO$_2$ capture have attracted much attention. For example, multi-N-containing ionic liquids can absorb much more SO$_2$/CO$_2$ in the flue gas than that of the limestone solvent [6]. Shiflett et al. [7,8] have found that the imidazolium-based ionic liquid [C$_4$mim][OAc] can markedly reduce the energy losses of CO$_2$ absorption comparing with those of the commercial monoethanolamine solvent. Guanidinium salt ILs (e.g., [TMG][L]) and functional ILs (e.g., [NH$_2$p-bmim][BF$_4$] and 2-(2-hydroxyethoxy) ammonium acetate) all show high efficiency for CO$_2$ and SO$_2$ capture [9,10].

The typical flue gas from coal-burning usually contains about 15 vol% CO$_2$, 10 vol% H$_2$O, and more than 2 vol% SO$_2$ [11]. Apart from CO$_2$, the effects of SO$_2$ on the flue gas absorption should be taken into account [12]. Most researchers consider that water has a little influence on the CO$_2$ capture, but the effects of SO$_2$ would not be neglected. For the impurities of the flue gas, it is found that the ILs are more likely to absorb SO_2 than CO_2. Specifically, the N element of the ionic liquid (IL) prefers to capture the SO_2 molecules, and then CO_2 molecules are repelled by the captured SO_2 [13]. Thus, the CO_2 absorption capacity of ionic liquids would be rapidly decreased over several cycles. Moreover, almost all of the ILs would exhibit much higher SO_2 absorption capacity than CO_2 due to the higher solubility of SO_2 in ILs. For instance, pure CO_2 solubility in the guanidinium-based ILs was only 0.4 mol/mol, while the SO_2 solubility in these ILs was as high as 2.5 mol/mol under the same conditions [10]. For an extreme case, the absorption capacities of pure SO_2 and CO_2 in the azole-based ILs (e.g., $[P_{66614}][Im]$) were 3.5 mol/mol and 0.1 mol/mol, respectively [14]. Although SO_2 has stronger interactions with ILs than CO_2 does, the actual partial pressure of SO_2 in the flue gas is very low [15]. It is well accepted that the partial pressure of SO_2 is about two orders of magnitude lower than that of CO_2, and low SO_2 partial pressure usually leads to low absorption capacity for SO_2. That is, a lot of energy will be consumed to remove SO_2 and CO_2 from the flue gas, if we used a two-step absorption process (absorbing SO_2 at first, and then CO_2).

The effects of single IL on the CO_2 capture have been widely discussed in recent years [2,7,16–19]. However, there are few reports on the IL mixtures for CO_2 absorption, especially on the CO_2 capture from the flue gas. Therefore, if the IL mixture was employed, the CO_2 absorption capacity of the mixed ILs might be greater than that of the single ionic liquid. Specifically, if the SO_2 of flue gas was absorbed by one IL of the mixture, the negative influence of SO_2 on the whole IL mixture might be significantly decreased, and then more CO_2 could be captured.

In this work, we want to use IL mixtures to remove CO_2 and SO_2 from the flue gas. The amine-functional ionic liquid $[NH_2emim][BF_4]$ and the imidazolium-based ionic liquid $[C_4mim][OAc]$ were synthesized at first. Subsequently, the two ILs were mixed to investigate the CO_2 and SO_2 absorption. Here, $[NH_2emim][BF_4]$ was mainly used to absorb SO_2 in the flue gas, while the $[C_4mim][OAc]$ ionic liquid was employed to capture CO_2. In order to more clearly study the actual flue gas, the CO_2 absorption performance in the IL mixture was measured at the simulated flue gas with 15 vol% CO_2 and 2 vol% SO_2. The effect of SO_2 on the IL mixture and the interaction between CO_2 and SO_2 in the IL absorption were also studied. Furthermore, the absorption mechanism at the molecule level was investigated by the quantum chemical calculation and the instrumental analysis.

2. Results and Discussion

2.1. CO_2 and SO_2 Absorption Performance of the Single ILs

When the simulated flue gas (15% CO_2/85% N_2) only contains the CO_2 impurity, one mole $[C_4mim][OAc]$ could absorb 0.298 mol CO_2 (Table 1). However, when SO_2 was mixed in the simulated flue gas (15% CO_2/2% SO_2/N_2), the CO_2 absorption capacity of $[C_4mim][OAc]$ was reduced to 0.204 CO_2/mol IL. The result shows that SO_2 has a negative effect on the IL absorbing CO_2. The single IL $[NH_2emim][OAc]$ and $[NH_2emim][BF_4]$ were also used to capture the CO_2 of flue gas. Without the interference of SO_2, the CO_2 absorption capacities of $[NH_2emim][OAc]$ and $[NH_2emim][BF_4]$ were 0.291 mol CO_2/mol IL and 0.290 mol CO_2/mol IL, respectively. However, like the $[C_4mim][OAc]$ ionic liquid, the CO_2 absorption capacities under exposure to 2% SO_2 would be markedly decreased to 0.171 mol CO_2/mol $[NH_2emim][OAc]$ and 0.180 mol CO_2/mol $[NH_2emim][BF_4]$, respectively. In short, the single ionic liquids all exhibit the CO_2 absorption capacities, but this capacity would be greatly weakened by the interference of SO_2.

Table 1. Summary of CO_2 absorption capacity by single ionic liquids.

Ionic Liquids	T, *K*	Gas	Absorption Capacity, mol CO_2/mol IL
[C$_4$mim][OAc]	293	15% CO_2/85% N_2	0.298
[C$_4$mim][OAc]	293	15% CO_2/2% SO_2/83% N_2	0.204
[NH$_2$emim][BF$_4$]	293	15% CO_2/85% N_2	0.290
[NH$_2$emim][BF$_4$]	293	15% CO_2/2% SO_2/83% N_2	0.180
[NH$_2$emim][OAc]	293	15% CO_2/85% N_2	0.291
[NH$_2$emim][OAc]	293	15% CO_2/2% SO_2/83% N_2	0.171

The researchers believed that CO_2 and SO_2 of the flue gas would be absorbed simultaneously [13,20]. The [C$_4$mim][OAc] and [NH$_2$emim][BF$_4$] absorption results supported this conclusion. For example, an experiment was carried out of outlet SO_2 concentration vs. time to investigate the SO_2 absorption performance. In this study (Figure 1a), the simulated flue gas contained 15% CO_2, 2% SO_2, and 83% N_2. The concentrations of CO_2 and SO_2 of the outlet stream were simultaneously detected with time. It was found that [C$_4$mim][OAc] and [NH$_2$emim][BF$_4$] all can capture CO_2, but the outlet SO_2 concentration was hardly detected before 40 min (Figure 1a). It indicates that SO_2 could be completely absorbed by [C$_4$mim][OAc] and [NH$_2$emim][BF$_4$] during the absorption process. Additionally, an extreme case was investigated in which the simulated flue gas contained 80% CO_2, 2% SO_2, and 18% N_2. However, SO_2 was also not found at the gas stream of the outlet before 15 min. Compared with CO_2, SO_2 has higher dipole moments and molecular polarity, which often results in the strong affinity of SO_2 with ionic liquids [21,22].

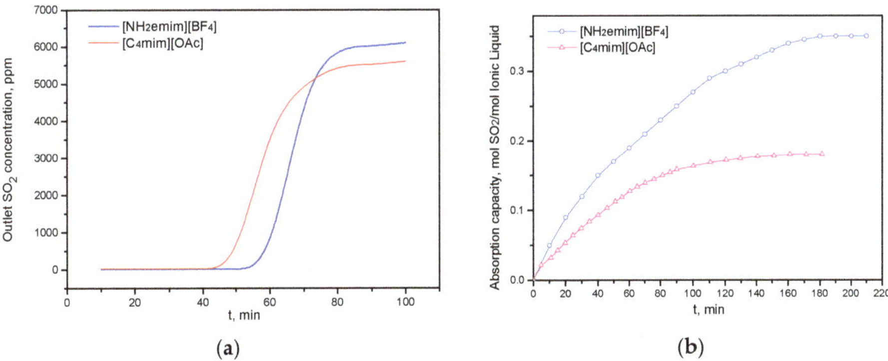

Figure 1. SO_2 absorption performance of [C$_4$mim][OAc] and [NH$_2$emim][BF$_4$]: (**a**) Outlet SO_2 concentration vs. time at the atmosphere of 15% CO_2, 2% SO_2, and 83% N_2; (**b**) SO_2 absorption capacity at the atmosphere of 2% SO_2/98% N_2.

The presence of SO_2 in flue gas usually leads to a competitive and negative influence on the separation of CO_2. Figure 2 shows the CO_2 absorption performance of [C$_4$mim][OAc] at the atmosphere of 15% CO_2/85% N_2 and 15% CO_2/2% SO_2/83% N_2, respectively. After 6 regeneration cycles, 1 mol [C$_4$mim][OAc] could absorb 0.255 mol CO_2, although the IL absorption capacity slightly decreased. In contrast, the absorption capacity of 1 mol [C$_4$mim][OAc] was only 0.10 mol CO_2 after the same number of cycles. In addition, the net SO_2 absorption experiment (2% SO_2/98% N_2) showed that [NH$_2$emim][BF$_4$] has high SO_2 absorption capacity (0.35 mol SO_2/mol IL), while 1 mol [C$_4$mim][OAc] only absorbs 0.18 mol SO_2 (Figure 1b). These results agreed well with previous studies [23–25].

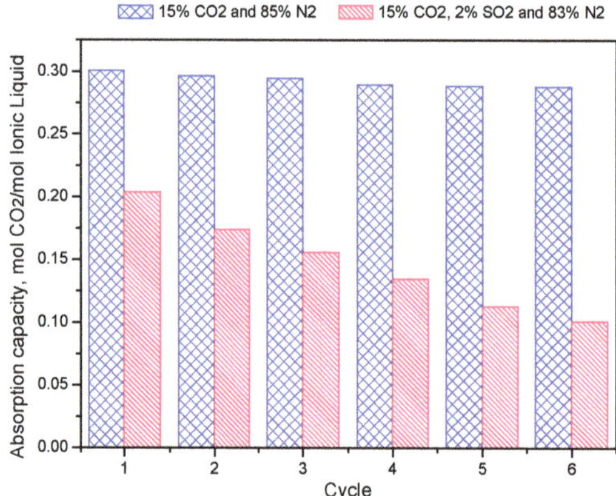

Figure 2. CO_2 absorption capacity of $[C_4mim][OAc]$ during 6 regeneration cycles at the atmosphere of 15% CO_2/85% N_2 and 15% CO_2/2% SO_2/83% N_2.

FT-IR can investigate the interaction between IL and CO_2/SO_2 [26]. The spectra of $[C_4mim][OAc]$ showed the changes after 15% CO_2 and 2% SO_2 absorption, respectively (Figure 3a–c). However, the $[C_4mim][OAc]$ spectrum had minimal changes when it was used to remove pure CO_2. Although the appearance of the carbonyl band at 1720 cm^{-1} shows that the acetate anion might be partly converted into the acetate acid [27], Shiflett considered that the amount of such a chemical reaction was minor and reversible [24]. Thus, the other reactions between CO_2 and the cation species in this work might not be detected within the wavenumber of 800–1600 cm^{-1}, as Figure 3a,c shown. Similarly, the FT-IR spectra of $[NH_2emim][BF_4]$ almost have no change before and after CO_2 absorption (Figure 3d,f), indicating that the chemical reaction between the $[NH_2emim]$ cation and CO_2 was not enough to be detected by FT-IR.

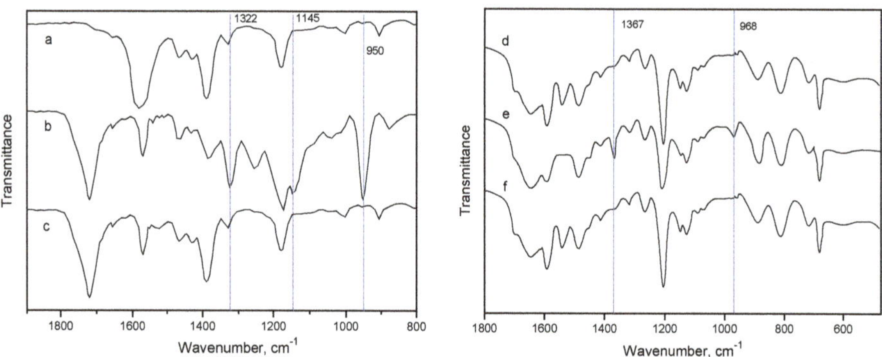

Figure 3. FT-IR spectra of ionic liquids (ILs): (**a**) Fresh $[C_4mim][OAc]$; (**b**) $[C_4mim][OAc]$ after SO_2 absorption (2% SO_2/98% N_2); (**c**) $[C_4mim][OAc]$ after CO_2 absorption (15% CO_2/85% N_2); (**d**) Fresh $[NH_2emim][BF_4]$; (**e**) $[NH_2emim][BF_4]$ after SO_2 absorption (2% SO_2/98% N_2); (**f**) $[NH_2emim][BF_4]$ after CO_2 absorption (15% CO_2/85% N_2).

In contrast, the chemical reaction between SO_2 and $[C_4mim][OAc]$ was much stronger than that of $[C_4mim][OAc]$–CO_2. Even if there was only a small amount of SO_2 (2%) in the flue gas, the FT-IR spectrum still shows marked changes (Figure 3b). Compared with the spectrum of fresh $[C_4mim][OAc]$, the peak intensity at 1580 and 1371 cm^{-1} decreased significantly as the $[C_4mim][OAc]$ absorbing SO_2. Meanwhile, the peaks at 1720, 1321, 1254, 1144, and 950 cm^{-1} newly appeared in the spectrum.

The bands at 1321 and 1144 cm^{-1} can be attributed to the stretching of SO_2 absorbed by the ionic liquid [26]. After SO_2 absorption, the new peak at 1720 cm^{-1} shows the formation of a carbonyl group, which also indicates that most of the acetate ions were no longer associated with [C$_4$mim] cations. The intense band at 950 cm^{-1} should be assigned to the vibrational mode of SO_3^{2-} or $S_2O_5^{2-}$. It once again suggests that the interaction between the SO_2 and [C$_4$mim][OAc] ionic liquid was strong. Similarly, the peak intensity at 885 and 1543 cm^{-1} changed markedly when the [NH$_2$emim][BF$_4$] absorbed SO_2. Particularly, two new peaks appeared at 968 and 1367 cm^{-1}, which can be attributed to the interaction between the N elements of the IL and SO_2 [6].

2.2. CO_2/SO_2 Absorption Properties in IL Mixtures

Due to the influence of SO_2, the CO_2 absorption capacity of single ionic liquids was greatly decreased. If [NH$_2$emim][BF$_4$] and [C$_4$mim][OAc] were simultaneously utilized, more CO_2 (with SO_2) in the flue gas might be captured. The CO_2 absorption capacities of IL mixtures at different mole fractions of [C$_4$mim][OAc] or [NH$_2$emim][BF$_4$] were displayed in Figure 4. Here, *X* is defined as the molar ratio of [NH$_2$emim][BF$_4$] to the IL mixture ([C$_4$mim][OAc]/[NH$_2$emim][BF$_4$]). Compared with the CO_2 absorption capacity of the single IL, the IL mixture could remove more CO_2. This result may be due to the presence of [NH$_2$emim][BF$_4$]. The small amount of SO_2 might be absorbed by [NH$_2$emim][BF$_4$] at first. Without the interference of SO_2, the CO_2 absorption capacity of the IL mixture was significantly enhanced. It was also found that the absorption of CO_2 did not increase significantly with the increase of the *X* value. When *X* was 0.3, the CO_2 absorption capacity of the [C$_4$mim][OAc]/[NH$_2$emim][BF$_4$] mixture reached up to the maximum. Similarly, when the IL mixture [C$_4$mim][OAc]/[NH$_2$emim][OAc] was used to remove CO_2/SO_2 of the flue gas, the poison effect of SO_2 on [C$_4$mim][OAc] was also greatly reduced. As Figure 4 shows, the CO_2 absorption of [C$_4$mim][OAc]/[NH$_2$emim][OAc] was about 0.4 mol CO_2/mol IL. Compared to the single [C$_4$mim][OAc], the absorption capacity of the IL mixture was improved.

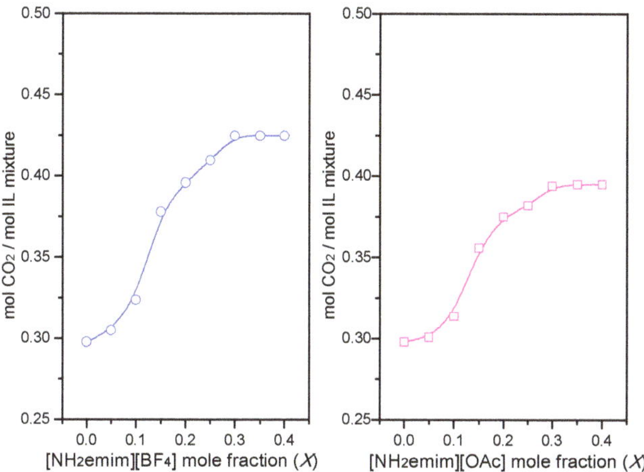

Figure 4. CO_2 absorption capacity of ionic liquid (IL) mixtures at different X: Flue atmosphere 15% CO_2/2% SO_2/83% N_2; absorption temperature, 293 K.

The ^{1}H NMR spectrum (Figure 5) shows that SO_2 would interact with [NH$_2$emim][BF$_4$] and [NH$_2$emim][OAc]. The NMR data of fresh [NH$_2$emim][BF$_4$] and fresh [NH$_2$emim][OAc] are listed as follows:

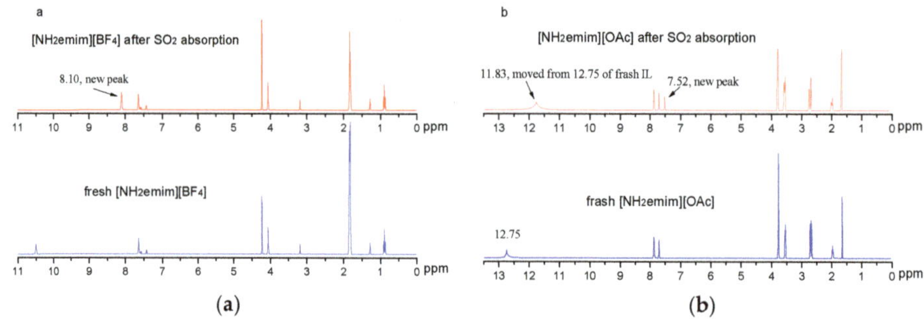

Figure 5. ^{1}H NMR spectra of ILs: (**a**) [NH$_2$emim][BF$_4$]; (**b**) [NH$_2$emim][OAc].

Fresh [NH$_2$emim][BF$_4$], δ = 7.721 (s, 1H, unsaturated C−H in the imidazole ring, with N connected to the left and right), 7.641 (d, 1H, unsaturated C−H in the imidazole ring), 7.636 (d, 1H, unsaturated C−H in the imidazole ring), 4.295 (s, 3H, H$_3$C−N ring), 4.039 (t, 2H, H$_2$C−N ring), 3.112 (m, 2H, N−CH$_2$−C−N ring), and 1.878 (t, 2H, NH$_2$).

Fresh [NH$_2$emim][OAc], δ = 12.751 (s, 1H, unsaturated C−H in the imidazole ring, with N connected to the left and right), 7.865 (d, 1H, unsaturated C−H in the imidazole ring), 7.703 (d, 1H, unsaturated C−H in the imidazole ring), 3.577 (s, 3H, H$_3$C−N ring), 3.749 (t, 2H, H$_2$C−N ring), 2.693 (m, 2H, N−CH$_2$−C−N ring), 1.973 (t, 2H, NH$_2$) and 1.651 (s, 3H, CH$_3$ in OAc-).

In comparison to the ^{1}H NMR spectrum of the fresh [NH$_2$emim][BF$_4$] (Figure 5a), new resonance peaks at 8.10 ppm were found after SO$_2$ absorption, which indicates the formation of S$\cdots$ N [28]. According to this result, it was considered that the interaction between SO$_2$ and [NH$_2$emim][BF$_4$] should mainly occur at the N element of the [NH$_2$emim] cation. For the case of the [NH$_2$emim][OAc] (Figure 5b), a typical peak of −COOH in the ^{1}H NMR spectrum moved from 12.75 to 11.83 ppm, and a new resonance peak was observed at 7.52 ppm after SO$_2$ absorption. These results suggest that the interaction between [NH$_2$emim][OAc] and SO$_2$ had occurred [25]. That is, the interaction between [OAc] and SO$_2$ leads to the moving from 12.75 ppm to 11.83, while the reaction of [NH$_2$emim] and SO$_2$ makes the new peak 7.52 ppm appearance.

In order to further investigate the effects of SO$_2$ on the CO$_2$ absorption capacity of ionic liquids, the CO$_2$ absorption performance of fresh IL and after SO$_2$-saturated IL are illustrated in Figure 6. Specifically, fresh [C$_4$mim][OAc] and fresh [NH$_2$emim][BF$_4$] ionic liquids were used to absorb SO$_2$ at first. When the IL was saturated by SO$_2$, the CO$_2$ absorption performance of the IL was investigated. It was found that the SO$_2$-saturated [C$_4$mim][OAc] did not have the ability to absorb CO$_2$. The concentration of CO$_2$ at the outlet was almost equal to that at the inlet. While fresh [C$_4$mim][OAc] can absorb CO$_2$ even after 60 min. The similar results were also observed using fresh [C$_2$mim][OAc] and SO$_2$ saturated [C$_2$mim][OAc] to absorb CO$_2$ [24]. Shiflet et al. considered that the interaction between the [OAc] anion and CO$_2$ plays an important role in the CO$_2$ removal of [OAc]−based ionic liquids [8]. However, the presence of SO$_2$ makes a great impact on the CO$_2$ absorption of [C$_4$mim][OAc]. In contrast, when [NH$_2$emim][BF$_4$] was saturated by SO$_2$, the [NH$_2$emim][BF$_4$] still had the CO$_2$ absorption capacity. As Figure 6b shows, SO$_2$-saturated [NH$_2$emim][BF$_4$] could capture about 2–7% CO$_2$ during the absorption process.

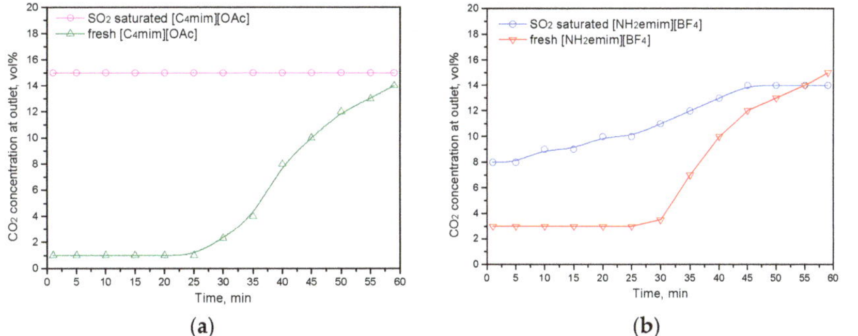

Figure 6. CO_2 absorption performance of fresh IL and after SO_2 saturated IL: (**a**) [C$_4$mim][OAc]; (**b**) [NH$_2$emim][BF$_4$].

2.3. Quantum Chemical Calculation on the Interaction of IL Mixture with CO_2/SO_2

The absorption capacity of CO_2 in the IL mixtures was higher than that of the single ionic liquid. This may be related to the interactions between ILs and CO_2/SO_2 molecules. Thus, the interaction of the [C$_4$mim][OAc] anion and [NH$_2$emim][BF$_4$] with CO_2/SO_2 was deeply investigated through quantum chemical calculation, which might be helpful to understand well the roles of CO_2 and SO_2 in the IL absorption. In this work, the structure of [NH$_2$emim][BF$_4$] and [C$_4$mim][OAc] was optimized on the basis of DFT-D3 calculation at first. The configuration of the IL with the lowest energy was considered as the optimized structure. Additionally, the structures of [C$_4$mim][OAc] and [NH$_2$emim][BF$_4$] with CO_2 and SO_2 were also investigated (Figure 7). The structural parameters for the IL−CO_2/SO_2 complexes are listed in Table 2.

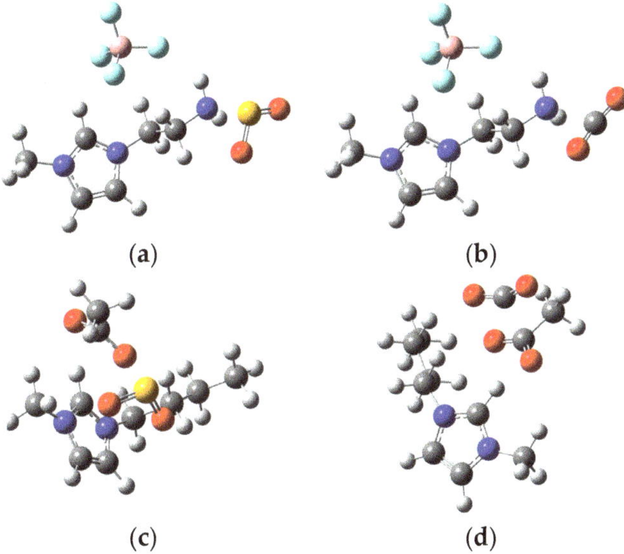

Figure 7. Optimized structure of the complexes of IL with CO_2/SO_2: (**a**) [NH$_2$emim][BF$_4$]−SO_2; (**b**) [NH$_2$emim][BF$_4$]−CO_2; (**c**) [C$_4$mim][OAc]−SO_2; (**d**) [C$_4$mim][OAc]−CO_2.

Table 2. Structural parameters for the complexes.

Structural Parameters	CO_2–[NH_2emim][BF_4]	SO_2–[NH_2emim][BF_4]	[C_4mim][OAc]–CO_2	[C_4mim][OAc]–SO_2
C–O, Å	1.19		1.21	
∠O–C–O, °	166		146	
S–O, Å		1.46		1.49
∠O–S–O, °		113.5		112.6

In general, the CO_2/SO_2 gas molecules around the anions and cations were related to the absorption reaction. As Figure 7a shows, there was a strong interaction between the N atom and the S atom of SO_2. Due to the complexation of $SO_2 \cdots$ N, the average angle of SO_2 was 116.0°. Compared to 119.5° of the pure SO_2 molecule, the bending degree of O=S=O was increased. Similarly, [NH_2emim][BF_4] also leads to an impact on the CO_2 structure. The angle of CO_2 was bent from 180° to 166°, and the bond length of C–O was extended from 1.16 Å to 1.19 Å. For the cases of [C_4mim][OAc], the interaction between the O and the C atom of carbon dioxide was also strong due to the negatively charged oxygen (O atom) in the [OAc] anion. The curvature of CO_2 could be increased by the interaction of the [OAc] anion. The average angle of CO_2 was bent to 146°, and the bond length of C–O was elongated to 1.21 Å. The configurations in Figure 7 also suggest that [NH_2emim] cation and [OAc] anion are the active sites for CO_2/SO_2 absorption.

To some extent, the interaction energy and absorption enthalpy might reflect the absorption capacity of the ILs. It was found that the NH_2emim] cation and [OAc] anion were the main active sites for the absorption of CO_2 and SO_2. In order to save the calculation cost and reduce the interference of other ions, herein, only the thermodynamic data of [OAc]–CO_2, [OAc]–SO_2, CO_2–[NH_2emim], and SO_2–[NH_2emim] complexes were compared (Table 3). It was found that the interaction energy and absorption enthalpy of [OAc]–CO_2–SO_2 complex were less than the sum of the energy and the enthalpy for [OAc]–CO_2 and [OAc]–SO_2, suggesting that CO_2 and SO_2 would competitively react with [OAc] anion.

Table 3. Thermochemical parameters and charge transfer of the ion–CO_2/SO_2 complexes.

	OAc–CO_2	OAc–SO_2	OAc–CO_2–SO_2
ΔE, kJ/mol	−40.7	−113.4	−140.0
ΔH, kJ/mol	−46.5	−125.1	−151.4
ΔG, kJ/mol	−1.6	−70.8	−72.1
net charge transfer, e	−0.510	−0.382	−0.035(CO_2)/−0.316(SO_2)
	CO_2–[NH_2emim]	SO_2–[NH_2emim]	SO_2–CO_2–[NH_2emim]
ΔE, kJ/mol	−33.8	−123.9	−156.8
ΔH, kJ/mol	−36.3	−126.7	−160.1
ΔG, kJ/mol	−7.2	−55.1	−61.9
net charge transfer, e	−0.312	−0.399	−0.308(CO_2)/−0.334(SO_2)

In contrast, the interaction energy and absorption enthalpy of [NH_2emim]–CO_2–SO_2 complex were approximately equal to the sum of those for the [NH_2emim]–CO_2 and [NH_2emim]–SO_2 complexes, indicating that the competitive reaction between [NH_2emim]–CO_2 and [NH_2emim]–SO_2 was not obvious. The absorption reaction might also lead to a change in charge distribution. It was found that the amount of net charge transfer from CO_2 to [NH_2emim] in [NH_2emim]–CO_2–SO_2 complex was almost equal to that of [NH_2emim]–CO_2, suggesting that [NH_2emim] had strong interactions with either SO_2 or CO_2. However, due to the impact of SO_2, the net charge transfer from [OAc] to CO_2 was significantly reduced from −0.510 in the [OAc]–CO_2 complex to −0.035 in [OAc]–CO_2–SO_2 complex, respectively, which might account for the decrease of CO_2 absorption capacity for [C_4mim][OAc] in Table 1.

Table 4 collects the thermochemical data of $[NH_2emim][BF_4]-CO_2$ and $[NH_2emim][BF_4]-SO_2$ complexes. In order to consider the solvent effect of ionic liquids, the continuum universal solvation model (SMD) was used in the calculation. Based on the SMD model, the interaction energies of $[NH_2emim][BF_4]-CO_2$ and $[NH_2emim][BF_4]-SO_2$ system were −16.8 and −76.3 kJ/mol, respectively. They were lower than those in the gas phase (−19.2 and −85.4 kJ/mol). Notably, the difference between the interaction energy ($[NH_2emim][BF_4]-CO_2$ and $[NH_2emim][BF_4]-SO_2$) in the gas phase (59.5 kJ/mol) was very consistent with the energy difference (66.2 kJ/mol) using the SMD model. In the liquid phase, the interaction energy between $[NH_2emim][BF_4]$ and SO_2 was slightly greater than that of $[NH_2emim][BF_4]-CO_2$, suggesting that $[NH_2emim][BF_4]$ tends to react with SO_2 rather than with CO_2 during the absorption process. Similarly, this phenomenon could also be observed in the thermodynamic data of the $[C_4mim][OAc]-CO_2$ and $[C_4mim][OAc]-SO_2$ complexes. In short, SO_2 was more active than CO_2 in the reaction with ionic liquids, and the $[NH_2emim][BF_4]$ may be more likely to absorb SO_2.

Table 4. Thermochemical parameters for the IL−CO_2/SO_2 complexes [a].

	$[C_4mim][OAc]-CO_2$	$[C_4mim][OAc]-SO_2$	$CO_2-[NH_2emim][BF_4]$	$SO_2-[NH_2emim][BF_4]$
ΔE, kJ/mol	−26.4 (−21.9)	−80.1 (−70.5)	−19.2 (−16.8)	−85.4 (−76.3)
ΔH, kJ/mol	−30.5	−93.2	−29.1	−91.5
ΔG, kJ/mol	−2.7	−40.6	8.1	−37.2

[a] Values of brackets were calculated by the continuum universal solvation model (SMD).

The interaction between IL mixture ($[C_4mim][OAc]$/$[NH_2emim][BF_4]$) and CO_2/SO_2 has also been investigated by the quantum chemistry calculation. The optimized structure is displayed in Figure 8. In the mixed ionic liquids, it is found that SO_2 was close to $[NH_2emim][BF_4]$, while the CO_2 molecule was near the $[C_4mim][OAc]$ ionic liquid. Specifically, SO_2 would have interacted with the N atom on the $[NH_2emim]$ cation, and CO_2 was more likely to react with the $[OAc]$ anion. This result might explain why the IL mixture can more effectively absorb the CO_2 of flue gas. Because of the existence of $[NH_2emim][BF_4]$, SO_2 may be first captured by $[NH_2emim]$. Without the interference of SO_2, the $[C_4mim][OAc]$ ionic liquid then could absorb more CO_2.

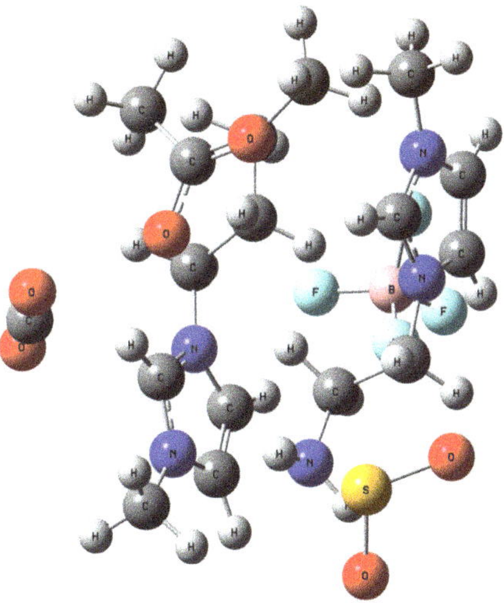

Figure 8. Optimized structure of the IL mixture and CO_2/SO_2.

3. Materials and Methods

3.1. Materials

The simulated flue gas was obtained by pure gas CO_2, SO_2, and N_2 (purity of >99.99 wt%). They were all purchased from Beifen (China) Gas Technology Company. 1-butyl-3-methylimidazolium tetrafluoroborate ([C_4mim][BF_4], 99 wt%) was obtained from Sigma-Aldrich Chemical Co., but the ionic liquid 1-butyl-3-methylimidazolium acetate ([C_4mim][OAc], 99 wt%) were purchased from Lanzhou Greenchem ILs, LICP, CAS, China. Additionally, [NH_2emim][BF_4] and [NH_2emim][OAc] have been synthesized by ourselves in this work. The used materials were as follows: 1-methylimidazole ($C_4H_6N_2$), 2-bromoethylamine hydrobromide ($C_2H_7Br_2N$), sodium acetate anhydrous (CH_3COONa), 1-methylimidazole ($C_4H_6N_2$), acetic acid (CH_3COOH), and sodium borate ($NaBF_4$). They were all provided by Sinopharm Chemical Reagent Co., Ltd., China, with purity over 98 wt%.

3.2. Ionic Liquid Preparation

In this work, the ionic liquids [NH_2emim][BF_4] and [NH_2emim][OAc] were prepared by ourselves according to the method used in the literature [17,18,29]. First, the [NH_2emim] cation was prepared by the reaction of 1-methylimidazole and 2-bromoethylamine hydrobromide under reflux for 12 h. Second, the [NH_2emim]-based IL was simply synthesized by ion exchange with $NaBF_4$ or $NaOAc/CH_3COOH$ in ethanol, and then the ethanol was removed in vacuum. The structures of the ILs were confirmed by proton nuclear magnetic resonance (^{1}H NMR, Bruker WB400 AMX spectrometer, Billerica, MA, USA). Here, deuterated chloroform ($CDCl_3$) was used as a solvent, and tetramethylsilane (TMS) was employed as an internal standard for ^{1}H NMR measurement.

3.3. CO_2 and SO_2 Absorption

As Figure 9 shows, CO_2 and SO_2 absorption experiments were performed in a 30 mL reactor immersed with a water-bath temperature controller. The temperatures were controlled at 293 K for absorption and 353 K for desorption, respectively. The simulated flue gas was a mixture of N_2, CO_2, and SO_2 in accordance with a certain proportion. As a typical absorption process, 10 mL IL or IL mixtures were added to the reactor at first. Subsequently, 15 vol% CO_2, 2 vol% SO_2 and 83 vol% N_2 were mixed in storage. The intake speed of the mixed gas was controlled at 60 mL/min, and the absorption pressure was controlled at 101.3 kPa. The concentrations of CO_2 and SO_2 were analyzed by a gas analyzer (MRU NOVA2000) at the outlet. To investigate the IL regeneration, CO_2 or SO_2 saturated IL was also loaded in the reactor. Desorption was performed at 353 K under a pure N_2 gas atmosphere for 30 min.

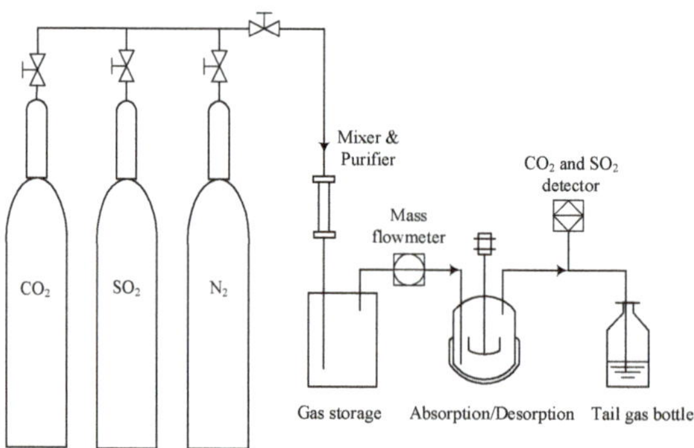

Figure 9. Schematic diagram of the apparatus used for absorption and desorption.

The amount of absorbed CO_2 or SO_2 was calculated by the following equation.

$$A_{gas} = \frac{M_{IL}\rho_{gas}Q\int_{t_1}^{t_2}(C_0 - C_{gas}(t))dt}{m_{IL}M_{gas}}$$

where A_{gas} is the molar amount of CO_2 or SO_2 in the ionic liquid; Q is the flow rate of the gas stream; C_0 and C_{gas} are the CO_2 or SO_2 concentrations at the inlet and the outlet streams, respectively; t_1 refers to the beginning time of the absorption process; when the CO_2 and SO_2 concentration at the outlet stream returns to the initial concentration, the time is t_2; M_{IL} and M_{gas} are the molecular weight of IL and CO_2 (or SO_2), respectively; m_{IL} is the weight of the ILs, and ρ_{gas} is the density of CO_2 or SO_2.

After the IL was saturated by CO_2 and SO_2, the complex structure was investigated by the FTIR and NMR technologies. The FTIR spectra of the samples were analyzed on an FTIR spectrometer (PerkinElmer, Frontier 2500). In addition, the structure changes of the [NH$_2$emim]-based IL after absorption were also detected by the NMR spectrometer (Bruker WB400 AMX, 300 MHz) using chloroform-*d* (CDCl$_3$) as a solvent and tetramethylsilane (TMS) as an internal standard.

3.4. Theoretical Calculation

A quantum chemical calculation was used in this work to study the interaction between the ionic liquid and CO_2 with SO_2. All calculations were carried out by the Gaussian 16 program [30]. For IL calculations, Li et al. [31] suggested that the density function of the Minnesota family [32] (e.g., M06-2X) with a diffusion function basis set (e.g., 6-311++G(d,p)) might give reasonable results. If dispersion-corrected density functionals (e.g., gd3bj, DFT-D3) were used, more reliable results could be obtained [33]. Therefore, the geometry optimization and frequency analysis of all ILs and IL mixtures were performed at the M06-2X/6-311++G(d,p) level and correction with Grimme's method. In order to calculate the interaction energy of the IL complexes, the basis set superposition error (BSSE) method was employed to correct the energy results [34]. The effect of the solvent should be taken into consideration in the theoretical calculation of the ionic liquids. It was found that the SMD solvation model proposed by Truhlar et al. can be used for the IL calculation very well [35,36]. Thus, the density functional theory (M06-2X and dispersion-corrected method) with the SMD model was also used to calculate the interaction energy of IL–CO_2/SO_2.

4. Conclusions

The CO_2 and SO_2 absorption of the flue gas in ionic liquids were investigated by the experimental method and theoretical calculation. The single ionic liquids, such as [NH$_2$emim][BF$_4$] and [C$_4$mim][OAc], all showed good CO_2 absorption performance for the simulated flue gas without SO_2 interference. However, SO_2 was more likely to react with the active sites of the ILs. When SO_2 was in the flue gas, the CO_2 absorption capacity of the single ionic liquid would be significantly inhibited. It was found that the interference of SO_2 on the CO_2 absorption performance might be markedly reduced by using the ionic liquid mixtures. The CO_2 absorption capacity of the IL mixture [C$_4$mim][OAc]/[NH$_2$emim][BF$_4$] was about 0.4 mol CO_2/mol IL even at an atmosphere of 15% CO_2/2%SO_2/83% N_2, which was greater than that of single [C$_4$mim][OAc] (0.204 mol CO_2/mol IL). There was a competitive relationship between CO_2 and SO_2 during the absorption process. The single ILs prefer to capture SO_2 rather than remove CO_2, due to the stronger interaction energy of SO_2 and the ILs. The experimental and calculated results suggested that the [OAc] anion and [NH$_2$emim] cation are the main active sites for CO_2 and SO_2 absorption. A lower absorption enthalpy of the IL–SO_2 or IL–CO_2 system usually means low absorption capacity. Thus, for the IL mixture [C$_4$mim][OAc]/[NH$_2$emim][BF$_4$], the quantum calculation results indicated that [NH$_2$emim][BF$_4$] might be more likely to absorb SO_2 of the flue gas and CO_2 was easily removed by the [C$_4$mim][OAc].

Author Contributions: Funding acquisition, Y.L.; investigation, G.W., Y.L., and G.L.; methodology, Y.L.; data curation, X.P.; writing—original draft, G.W., and Y.L. All authors have read and agreed to the published version of the manuscript.

Funding: This research and the APC were funded by the National Natural Science Foundation of China (grant numbers: 21766021 and 21266015).

Conflicts of Interest: The authors declare no conflict of interest.

References

1. Aghaie, M.; Rezaei, N.; Zendehboudi, S. A systematic review on CO_2 capture with ionic liquids: Current status and future prospects. *Renew. Sustain. Energy Rev.* **2018**, *96*, 502–525. [CrossRef]
2. Yu, C.H.; Huang, C.H.; Tan, C.S. A review of CO_2 capture by absorption and adsorption. *Aerosol Air Qual. Res.* **2012**, *12*, 745–769. [CrossRef]
3. Wang, B.; Qin, L.; Mu, T.; Xue, Z.; Gao, G. Are ionic liquids chemically stable? *Chem. Rev.* **2017**, *117*, 7113–7131. [CrossRef]
4. Zhang, S.; Sun, N.; He, X.; Lu, X.; Zhang, X. Physical properties of ionic liquids: Database and evaluation. *J. Phys. Chem. Ref. Data* **2006**, *35*, 1475–1517. [CrossRef]
5. Blanchard, L.A.; Hancu, D.; Beckman, E.J.; Brennecke, J.F. Green processing using ionic liquids and CO_2. *Nature* **1999**, *399*, 28–29. [CrossRef]
6. Wang, C.M.; Cui, G.K.; Luo, X.Y.; Xu, Y.J.; Li, H.R.; Dai, S. Highly efficient and reversible SO_2 capture by tunable azole-based ionic liquids through multiple-site chemical absorption. *J. Am. Chem. Soc.* **2011**, *133*, 11916–11919. [CrossRef]
7. Shiflett, M.B.; Yokozeki, A. Separation of carbon dioxide and sulfur dioxide using room-temperature ionic liquid [bmim][MeSO4]. *Energy Fuels* **2010**, *24*, 1001–1008. [CrossRef]
8. Shiflett, M.B.; Yokozeki, A. Solubilities and diffusivities of carbon dioxide in ionic liquids: Bmim PF_6 and bmim BF_4. *Ind. Eng. Chem. Res.* **2005**, *44*, 4453–4464. [CrossRef]
9. Zhang, X.; Jiang, K.; Liu, Z.; Yao, X.; Liu, X.; Zeng, S.; Dong, K.; Zhang, S. Insight into the performance of acid gas in ionic liquids by molecular simulation. *Ind. Eng. Chem. Res.* **2019**, *58*, 1443–1453. [CrossRef]
10. Shang, Y.; Li, H.P.; Zhang, S.J.; Xu, H.; Wang, Z.X.; Zhang, L.; Zhang, J.M. Guanidinium-based ionic liquids for sulfur dioxide sorption. *Chem. Eng. J.* **2011**, *175*, 324–329. [CrossRef]
11. Izgorodina, E.I.; Hodgson, J.L.; Weis, D.C.; Pas, S.J.; MacFarlane, D.R. Physical absorption of CO_2 in protic and aprotic ionic liquids: An interaction perspective. *J. Phys. Chem. B* **2015**, *119*, 11748–11759. [CrossRef] [PubMed]
12. Karousos, D.S.; Kouvelos, E.; Sapalidis, A.; Pohako-Esko, K.; Bahlmann, M.; Schulz, P.S.; Wasserscheid, P.; Siranidi, E.; Vangeli, O.; Falaras, P.; et al. Novel inverse supported ionic liquid absorbents for acidic gas removal from flue gas. *Ind. Eng. Chem. Res.* **2016**, *55*, 5748–5762. [CrossRef]
13. Li, X.S.; Zhang, L.Q.; Zheng, Y.; Zheng, C.G. Effect of SO_2 on CO_2 absorption in flue gas by ionic liquid 1-ethyl-3-methylimidazolium acetate. *Ind. Eng. Chem. Res.* **2015**, *54*, 8569–8578. [CrossRef]
14. Wang, C.; Luo, X.; Luo, H.; Jiang, D.E.; Li, H.; Dai, S. Tuning the basicity of ionic liquids for equimolar CO_2 capture. *Angew. Chem. Int. Ed.* **2011**, *50*, 4918–4922. [CrossRef] [PubMed]
15. Lei, Z.G.; Dai, C.N.; Chen, B.H. Gas solubility in ionic liquids. *Chem. Rev.* **2014**, *114*, 1289–1326. [CrossRef] [PubMed]
16. Goodrich, B.F.; de la Fuente, J.C.; Gurkan, B.E.; Zadigian, D.J.; Price, E.A.; Huang, Y.; Brennecke, J.F. Experimental measurements of amine-functionalized anion-tethered ionic liquids with carbon dioxide. *Ind. Eng. Chem. Res.* **2011**, *50*, 111–118. [CrossRef]
17. Bates, E.D.; Mayton, R.D.; Ntai, I.; Davis, J.H. CO_2 capture by a task-specific ionic liquid. *J. Am. Chem. Soc.* **2002**, *124*, 926–927. [CrossRef]
18. Wang, M.; Zhang, L.Q.; Gao, L.X.; Pi, K.W.; Zhang, J.Y.; Zheng, C.G. Improvement of the CO_2 absorption performance using ionic liquid [NH_2emim][BF_4] and [Emim][BF_4]/[Bmim][BF_4] mixtures. *Energy Fuels* **2013**, *27*, 461–466. [CrossRef]
19. Damas, G.B.; Dias, A.B.A.; Costa, L.T. A quantum chemistry study for ionic liquids applied to gas capture and separation. *J. Phys. Chem. B* **2014**, *118*, 9046–9064. [CrossRef]
20. Amarasekara, A.S. Acidic Ionic Liquids. *Chem. Rev.* **2016**, *116*, 6133–6183. [CrossRef]

21. Rezaei, F.; Jones, C.W. Stability of supported amine adsorbents to SO_2 and NO_x in postcombustion CO_2 capture. 1. Single-component adsorption. *Ind. Eng. Chem. Res.* **2013**, *52*, 12192–12201. [CrossRef]

22. Hallenbeck, A.P.; Kitchin, J.R. Effects of O_2 and SO_2 on the capture capacity of a primary-amine based polymeric CO_2 sorbent. *Ind. Eng. Chem. Res.* **2013**, *52*, 10788–10794. [CrossRef]

23. Shiflett, M.B.; Yokozeki, A. Chemical absorption of sulfur dioxide in room-temperature ionic liquids. *Ind. Eng. Chem. Res.* **2010**, *49*, 1370–1377. [CrossRef]

24. Shiflett, M.B.; Drew, D.W.; Cantini, R.A.; Yokozeki, A. Carbon dioxide capture using ionic liquid 1-butyl-3-methylimidazolium acetate. *Energy Fuels* **2010**, *24*, 5781–5789. [CrossRef]

25. Li, W.; Liu, Y.; Wang, L.; Gao, G. Using ionic liquid mixtures to improve the so_2 absorption performance in flue gas. *Energy Fuels* **2017**, *31*, 1771–1777. [CrossRef]

26. Lee, K.Y.; Kim, H.S.; Kim, C.S.; Jung, K.D. Behaviors of SO_2 absorption in BMIm OAc as an absorbent to recover SO_2 in thermochemical processes to produce hydrogen. *Int. J. Hydrog. Energy* **2010**, *35*, 10173–10178. [CrossRef]

27. Gurau, G.; Rodríguez, H.; Kelley, S.P.; Janiczek, P.; Kalb, R.S.; Rogers, R.D. Demonstration of chemisorption of carbon dioxide in 1,3-dialkylimidazolium acetate ionic liquids. *Angew. Chem. Int. Ed.* **2011**, *50*, 12024–12026. [CrossRef] [PubMed]

28. Zhai, L.Z.; Zhong, Q.; He, C.; Wang, J. Hydroxyl ammonium ionic liquids synthesized by water-bath microwave: Synthesis and desulfurization. *J. Hazard. Mater.* **2010**, *177*, 807–813. [CrossRef]

29. Kurnia, K.A.; Mutalib, M.I.A.; Ariwahjoedi, B. Estimation of physicochemical properties of ionic liquids [H_2N-C_2mim][BF_4] and [H_2N-C_3mim][BF_4]. *J. Chem. Eng. Data* **2011**, *56*, 2557–2562. [CrossRef]

30. Frisch, M.J.; Trucks, G.W.; Schlegel, H.B.; Scuseria, G.E.; Robb, M.A.; Cheeseman, J.R.; Scalmani, G.; Barone, V.; Mennucci, B.; Petersson, G.A.; et al. *Gaussian 16 (Revision A.01)*; Gaussian, Inc.: Wallingford, CT, USA, 2016.

31. Li, H.P.; Chang, Y.H.; Zhu, W.S.; Jiang, W.; Zhang, M.; Xia, J.X.; Yin, S.; Li, H.M. A DFT Study of the Extractive Desulfurization Mechanism by BMIM (+) AlCl4 (-) Ionic Liquid. *J. Phys. Chem. B* **2015**, *119*, 5995–6009. [CrossRef]

32. Zhao, Y.; Truhlar, D.G. A new local density functional for main-group thermochemistry, transition metal bonding, thermochemical kinetics, and noncovalent interactions. *J. Chem. Phys.* **2006**, *125*, 194101–194118. [CrossRef] [PubMed]

33. Grimme, S.; Hujo, W.; Kirchner, B. Performance of dispersion-corrected density functional theory for the interactions in ionic liquids. *Phys. Chem. Chem. Phys.* **2012**, *14*, 4875–4883. [CrossRef] [PubMed]

34. Boys, S.F.; Bernardi, F. The Calculations of Small Molecular Interaction by the Difference of Separate Total Energies. Some Procedures with Reduced Error. *Mol. Phys.* **1970**, *19*, 553–566. [CrossRef]

35. Bernales, V.S.; Marenich, A.V.; Contreras, R.; Cramer, C.J.; Truhlar, D.G. Quantum Mechanical Continuum Solvation Models for Ionic Liquids. *J. Phys. Chem. B* **2012**, *116*, 9122–9129. [CrossRef] [PubMed]

36. Marenich, A.V.; Cramer, C.J.; Truhlar, D.G. Universal solvation model based on solute electron density and on a continuum model of the solvent defined by the bulk dielectric constant and atomic surface tensions. *J. Phys. Chem. B* **2009**, *113*, 6378–6396. [CrossRef]

Sample Availability: Samples of the ionic liquids are available from the authors.

molecules

Article

Studies on the Solubility of Terephthalic Acid in Ionic Liquids

Karolina Matuszek [1], Ewa Pankalla [2], Aleksander Grymel [2], Piotr Latos [3] and Anna Chrobok [3,*]

[1] Monash University, School of Chemistry, Clayton, VIC 3800, Australia; karolina.matuszek@monash.edu
[2] Grupa Azoty Zakłady Azotowe Kędzierzyn, S.A., Mostowa 30A, 47-220 Kędzierzyn-Koźle, Poland; ewa.pankalla@zak.com.pl (E.P.); ALEKSANDER.GRYMEL@GRUPAAZOTY.COM (A.G.)
[3] Department of Chemical Organic Technology and Petrochemistry, Silesian University of Technology, Krzywoustego 4, 44-100 Gliwice, Poland; Piotr.Latos@polsl.pl
* Correspondence: Anna.Chrobok@polsl.pl; Tel.: +48-32-237-1032

Academic Editors: Monica Nardi, Antonio Procopio and Maria Luisa Di Gioia
Received: 26 November 2019; Accepted: 23 December 2019; Published: 24 December 2019

Abstract: Low solubility of terephthalic acid in common solvents makes its industrial production very difficult and not environmentally benign. Ionic liquids are known for their extraordinary solvent properties, with capability to dissolve a wide variety of materials, from common solvents to cellulose, opening new possibilities to find more suitable solvents for terephthalic acid. This work presents studies on the solubility of terephthalic acid in ionic liquids, and demonstrates that terephthalic acid is soluble in ionic liquids, such as 1-ethyl-3-methylimidazolium diethylphosphate, 1-butyl-3-methylimidazolium acetate, and dialkylimidazolium chlorides up to four times higher than in DMSO. Additionally, the temperature effect and correlation of ionic liquid structure with solubility efficiency are discussed.

Keywords: terephthalic acid; ionic liquids; solubility

---

1. Introduction

Purified terephthalic acid (PTA) is a white, crystalline solid with negligible vapor pressure under standard conditions [1]. Nearly all PTA is consumed in polyester production, including polyester fiber and film, and polyethylene terephthalate (PET) resin. Nowadays, PTA is produced by the catalytic liquid-phase oxidation of p-xylene in acetic acid, in the presence of air, using a manganese or cobalt acetate catalyst. In 2011, global PTA capacity reached 28.8 million tons, with China being the main world producer [2].

Terephthalic acid is poorly soluble in organic solvents (Table 1) [1,3]. Among all tested solvents the best solubility of PTA was observed in DMSO (20 g of PTA per 100 g DMSO at 25 °C). PTA is also soluble in N-methyl-2-pyrolidone and dimethylamine, but the solubility at 90 °C is two times lower than in DMSO [3]. In general, the solubility of PTA in organic solvents is low and slowly increases with increasing temperature.

Low solubility of PTA in conventional solvents creates problems in industry during the purification and transformation of PTA into useful chemicals. The crude terephthalic acid (CTA) contains impurities like p-toluic acid and 4-carboxybenzaldehyde (CBA). CTA is purified by crystallization from water, dissolving CTA at 300 °C and elevated pressure; however, under these conditions the solubility is only 40 mass%. As a result, this methodology is expensive and energy consuming. This problem also arises in the other applications like chemical transformations, where low miscibility of PTA with other reagents results in low product rate. One of the examples is the production of esters of terephthalic acid as alternative plasticizers. Poor solubility of PTA in alcohols forces the use of higher pressures and temperatures in the esterification processes.

Table 1. Solubility of PTA (terephthalic acid) in organic solvents (g_{PTA}/100 g solvent) [1].

Solvent	Temperature		
	25 °C	150 °C	200 °C
Water	0.0017	0.24	1.7
Methanol	0.1	3.1	lack of data
DMF	7.4	lack of data	lack of data
DMSO	20.0	lack of data	lack of data

DMF, dimethylformamide; DMSO, dimethyl sulfoxide.

Therefore, alternative solvents, which can overcome dissolution limitations, are of high importance for the processing of PTA. Ionic liquids (ILs) with their unique properties can help to solve some of the described problems above [4–8]. Ionic liquids make promising alternatives to conventional solvents, and lead to a both greener and economically viable process. Ionic liquids as sustainable solvents play an important role in pharmacological development [9,10] and in the chemical industry [11]. Various approaches of implementing ionic liquids in organic synthesis as solvents and catalysts were also demonstrated [12,13].

Literature concerning solubilities of PTA in ionic liquids is scarce. However, several patents concerning purification of the CTA from by-products using ionic liquids exists, and solubility data are presented [14–16]. Among them is a patent concerning the purification of aryl carboxylic acids published in 2010, which reports that the impure acid is dissolved or dispersed in an ionic liquid, which is followed by the addition of a non-solvent to precipitate the acid while other impurities remain dissolved. In this case, the term "non-solvent" defines a compound that is highly soluble in ionic liquids with little or no solubility in aryl carboxylic acid [14]. A detailed graph comparing the solubility of PTA in ionic liquids and conventional solvents was provided (Figure 1) [5]. High solubility of PTA in several ionic liquids was shown in relatively low temperatures. According to this data, PTA can be dissolved in ionic liquids to a greater extent than in solvents such as DMSO, DMF, or water. Among all studied ionic liquids, the solubility of PTA was shown in the following order: 1-ethyl-3-methylimidazolium diethylphosphate [C$_2$mim][Et$_2$PO$_4$] > 1-butyl-3-methylimidazolium chloride [C$_4$mim]Cl > 1-ethyl-3-methylimidazolium chloride [C$_2$mim]Cl > *N*-ethyl-2-methylpyridinium ethyl sulfate [C$_2$mpy][EtSO$_4$] > 1-ethyl-3-methylimidazolium ethyl sulfate [C$_2$mim][EtSO$_4$]. For example, 37 mass% of PTA was dissolved in [C$_2$mim][Et$_2$PO$_4$] at 100 °C, while in DMSO only 15 mass% could be dissolved.

The next patent, published in 2013, describes purification of CTA containing CBA as impurity with ionic liquids [1]. After 2 h, the mixture of CTA and an ionic liquid was cooled down to room temperature and the solid (PTA) was filtered off. The best results were obtained with the application of trihexyl(tetradecyl)phosphine bromide (Cyphos 102), trihexyl(tetradecyl)phosphine chloride (Cyphos 101), 1-butyl-1-methylpyrrolidinium bis(trifluoromethanesulfonyl)imide ([bmpyr][NTf$_2$]) and 1-butyl-3-methylimidazolium acetate ([bmim][OAc]). Application of these ionic liquids led to the removal of 99%, 97%, 93%, and 90% of CBA, respectively. Information about the exact amounts of PTA dissolved in ionic liquids was not presented.

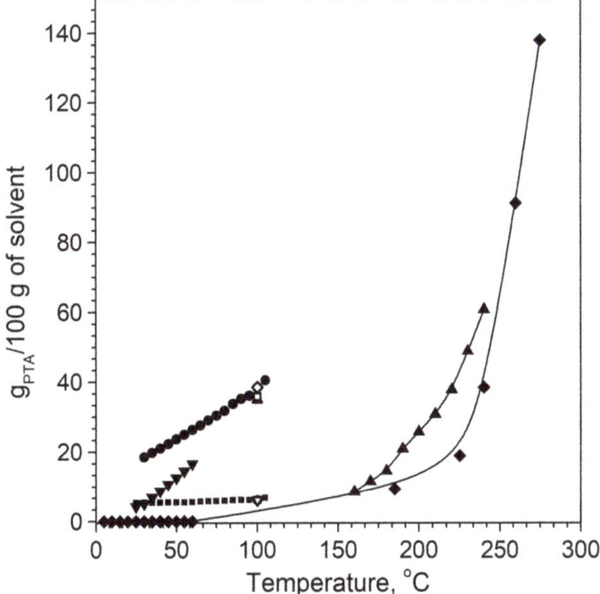

Figure 1. Solubility of PTA in ionic liquids and conventional solvents; figure adapted from Rogers, R. D et al. US 2010174111 A1, 20109 (◇) [C$_2$mim][Et2PO4], (Δ) [C$_4$mim]Cl, (□) [C$_2$mim]Cl, (∇) [C$_2$mpy][EtSO$_4$], (○) [C$_2$mim][EtSO$_4$], (♦) water, (*K*) acetic acid, (■) DMF, (*L*)N-Methyl-2-pyrrolidone, (●) DMSO.

In 2015, a patent was issued that involved the separation of aryl carboxylic acids from a mixture comprising at least two acids [16]. The separation process is based on heating the mixture of at least two aryl carboxylic acids with an ionic liquid, and cooling down using fractional crystallization to precipitate the desired acid. Information about solubility of PTA in ionic liquids can be found in several examples in the patent: A total of 59.28 g of PTA/100 g of choline chloride at 224 °C, 47.56 g of PTA/100g of 1-butyl-3-methylimidazolium chloride at 160 °C, 6.25 g of PTA/100g of trihexyltetradecylphosphonium bromide at 160 °C, 45.10 g of PTA/100 g of choline bromide at 220 °C, 37.89 g of PTA/100 g of 1-butyl-3-methylimidazolium bromide at 180 °C, 17.54 g of PTA/100 g of 1-butyl-3-methylimidazolium methanesulfonate at 120 °C.

In summary, patents, which describe purification of CTA from typical impurities, such as CBA or p-toluic acid, provide some data concerning the solubility of PTA in ionic liquids. This data shows that the solubility of PTA in ionic liquids is higher than in conventional solvents; however, in many cases there is a lack of detailed information concerning the amounts of PTA dissolved in ionic liquids at a given temperature.

This work provides detailed research concerning the solubility of terephthalic acid in ionic liquids as a function of temperature, and also discusses the influence of the cation and anion structure on the solubility of PTA in various ionic liquids.

2. Results and Discussion

Ionic liquids used in this work were chosen based on the patent literature described above [9–11]. Solubility studies were conducted at temperatures between 25 and 100 °C and the obtained results are summarized in Table 2 in order of decreasing solubility.

Table 2. Solubility of PTA in ionic liquids.

No.	Ionic Liquid	PTA Solubility g/100 g IL ($\pm$0.5 g)				
		25 °C	40 °C	60 °C	80 °C	100 °C
1	1-ethyl-3-methylimidazolium diethylphosphate [C$_2$mim][Et$_2$PO$_4$]	52.2	59.0	60.3	62.1	63.4
2	1-butyl-3-methylimidazolium acetate [C$_4$mim][OAc]	6.0	15.1	28.6	36.4	57.1
3	1-ethyl-3-methylimidazolium chloride [C$_2$mim]Cl	solid	solid	solid	38.8	42.9
4	1-butyl-3-methylimidazolium chloride [C$_4$mim]Cl	solid	solid	27.2	32.1	34.0
5	1-hexyl-3-methylimidazolium chloride [C$_6$mim]Cl	h.v.	13.3	27.9	32.4	34.2
6	1-octyl-3-methylimidazolium chloride [C$_8$mim]Cl	h.v.	h.v.	5.6	15.6	30.9
7	DMSO	20.5	23.5	25.6	27.9	29.4
8	1-butyl-3-methylimidazolium dicyanamide [C$_4$mim][N(CN)$_2$]	0.8	1.6	2.3	2.9	21.8
9	1-butyl-1-methylpyridinum chloride [bmpyr]Cl	solid	solid	solid	solid	25.9
10	1-methylimidazolium acetate [Hmim][OAc]	0	3.0	6.8	10.7	24.7
11	1-butyl-1-methylpyrrolidinium chloride [bmpyrr]Cl	solid	solid	solid	solid	17.4
12	Tetrabutylammonium bromide[N$_{4,4,4,4}$]Br	solid	solid	solid	solid	19.1
13	Tetradecyl(trihexyl)phosphonium chloride [P$_{14,6,6,6}$]Cl	0	4.4	6.3	7.9	13.2
14	Tetrabutylammonium chloride [N$_{4,4,4,4}$]Cl	solid	solid	solid	solid	6.7
15	1-methylimidazolium chloride [Hmim]Cl	solid	solid	2.0	4.4	9.5
16	1-butyl-3-methylimidazolium methylsulfate [C$_4$mim][MeSO$_4$]	0	2.3	4.0	5.2	7.6
17	1-ethyl-3-methylimidazolium ethylsulfate [C$_2$mim][EtSO$_4$]	0	0	0	4.7	5.9
18	Tetradecyl(trihexyl)phosphonium bromide [P$_{14,6,6,6}$]Br	0	0	1.4	2.8	5.2
19	1-butyl-3-methylimidazolium hydrogensulfate [C$_4$mim][HSO$_4$]	h.v.	0	0	0	1.7
20	1-butyl-3-methylimidazolium bis(trifluoromethylsulfonyl)imide [C$_4$mim][NTf$_2$]	0	0	0	0	0
21	1-butyl-3-methylimidazolium trifluoromethanesulfonate [C$_4$mim][OTf]	0	0	0	0	0
22	1-butyl-3-methylimidazolium hexafluorophosphate [C$_4$mim][PF$_6$]	0	0	0	0	0
23	1-butyl-3-methylimidazolium tetrafluoroborate [C$_4$mim][BF$_4$]	0	0	0	0	0
24	1-methylimidazolium hydrogen sulfate [Hmim][HSO$_4$]	0	0	0	0	0

h.v., high viscosity at a given temperature.

Among the studied ionic liquids, the best results were obtained for [C$_2$mim][Et$_2$PO$_4$]. At room temperature, 52.2 g$_{PTA}$/100 g$_{IL}$ was dissolved, and 63.4 g$_{PTA}$/100 g$_{IL}$ at 100 °C, which is two times greater than in DMSO (29.4 g_{PTA}/100 g_{DMSO} at 100 °C). High solubilities were also observed for [C$_4$mim][OAc] (57.1 g_{PTA}/100 g_{IL}, 100 °C) and the dialkylimidazolium chlorides [C$_2$mim]Cl (42.9 g_{PTA}/100 g_{IL}, 100 °C), [C$_4$mim]Cl (34.0 g_{PTA}/100 g_{IL}, 100 °C) and [C$_6$mim]Cl (34.2 g_{PTA}/100 g_{IL}, 100 °C). All other tested ionic liquids exhibited lower solubility than DMSO. It can be concluded that the highest solubilities of PTA were found in ionic liquids that were aprotic, contained Lewis base properties (similar to DMSO [17]), and possessed chloride, acetate, or diethylphosphate anions. Protic ionic liquids with Lewis base properties ([Hmim][OAc]) also show relatively high PTA solubility (24.7 g_{PTA}/100 g_{IL}, 100 °C); although, this is lower than its aprotic homologue ([C$_4$mim][OAc]), most probably due to the weak Brønsted acid properties of the free proton on the nitrogen atom in the [Hmim]$^+$ cation.

The solubility of terephthalic acid in ionic liquids slowly increases with temperature (52.2 g_{PTA}/100 g_{IL} at 25 °C, and 63.4 g_{PTA}/100 g_{IL} at 100 °C). This effect is much more pronounced in the case of conventional solvents (0.1 g_{PTA}/100 g_{MeOH} at 25 °C, and 3.1 g_{PTA}/100 g_{MeOH} at 150 °C). This can be explained by the solubility in conventional solvents at room temperature being relatively low, with only the increase of the average kinetic energy, caused by increase of temperature, allowing the solvent molecules to overcome intermolecular attraction and break apart the terephthalic acid molecules. On the other hand, some ionic liquids can already overcome those attraction forces at room temperature and; therefore, temperature only slightly increases the solubility.

2.1. The Influence of the Structure of Cation in Ionic Liquids

It was found that the structure of cation in the ionic liquid affects the PTA solubility. Four ionic liquids with the same chloride anion and with different structures of the cation were compared in Figure 2. Tested cations represent a variety of structures, such as aromatic—1-butyl-4-methylpiridinium and 1-ethyl-3-methylimidazolium, alicyclic—1-butyl-1-methylpyrrolidinium and aliphatic—tetrabutylammonium. Ionic liquids with aliphatic substituents in the ammonium cation exhibit the lowest solubility of PTA, only 6.7 g_{PTA}/100 g_{IL} at 100 °C. The use of a cyclic, non-aromatic cation allows for slightly better results: 17.4 g_{PTA}/100 g_{IL} at 100 °C. Ionic liquids with aromatic cations turned out to be the most effective (42.9 g_{PTA}/100 g_{IL}). Most likely, this is due to the effect of π–π interactions between aromatic rings located in both ionic liquid cation and PTA [18].

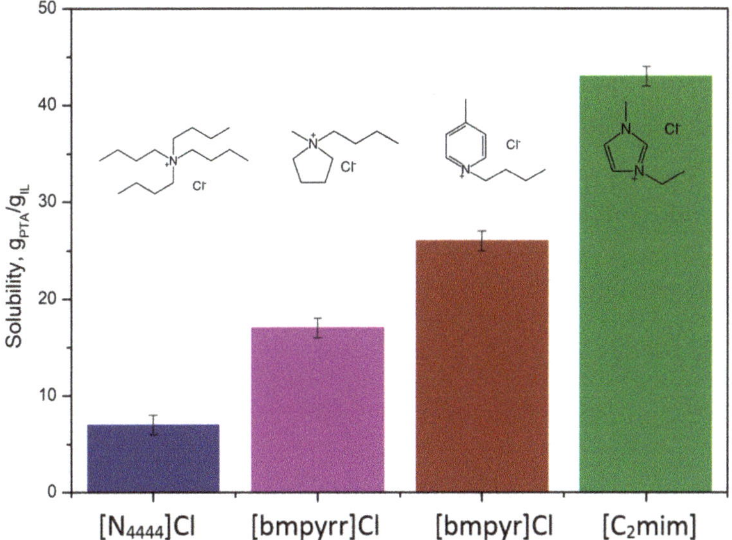

Figure 2. The influence of the structure of cation in ionic liquids on the solubility of PTA at 100 °C.

2.2. The Influence of the Alkyl Side Chain Length in the Cation

The dialkylimidazolium cation possesses two alkyl substituents in the structure, which influence the physicochemical properties of ionic liquid (e.g., density, melting point, and viscosity). Ionic liquids based on the 1-alkyl-3-methylimidazium cation with various side alkyl chain length—ethyl $[C_2mim]^+$, butyl $[C_4mim]^+$, hexyl $[C_6mim]^+$, octyl $[C_8mim]^+$—and also an ionic liquid with a proton instead of an alkyl chain $[Hmim]^+$, were selected for these studies. Obtained results are presented in Figure 3. The first three homologs are crystalline solids at room temperature. The elongation of the alkyl chain to six carbons causes a symmetry disorder and, consequently, $[C_6mim]Cl$ and $[C_8mim]Cl$ are very viscous liquids, with a viscosity of 715 and 337 cP at 25 °C, respectively. The PTA solubility decreases as the length of the alkyl chain increases from ethyl to butyl or octyl. Similar observations were noted for ILs and water [19]. Better solubility of PTA in the 1-ethyl-3-methylimidazolium chloride can be attributed to the higher charge density and polarity, compared to the longer homologues.

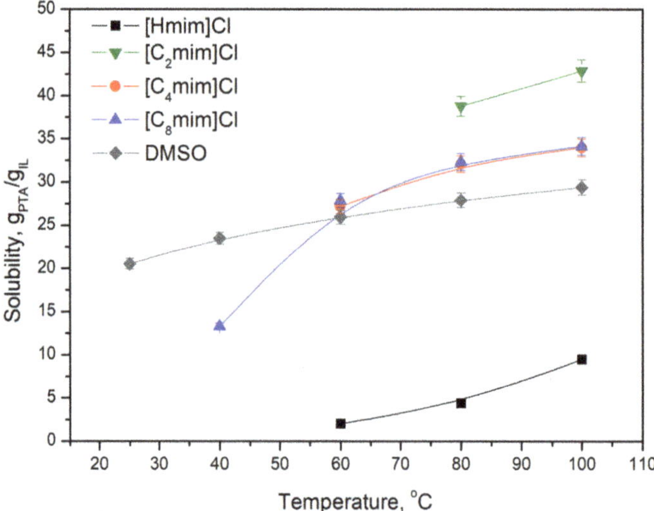

Figure 3. The influence of the alkyl chain length in the 1-alkyl-3-methylimidazolium cation on the solubility of PTA.

2.3. The Influence of the Structure of Anion in Ionic Liquids

To observe the influence of the anion on the solubility of PTA, a wide range of ionic liquids with the same 1-methyl-3-butylimidazolium cation and different anions were studied. The anions showing neutral (in the acid/base sense: $[BF_4]^-$, $[PF_6]^-$, $[CH_3SO_3]^-$, $[NTf_2]^-$), acidic ($[OTf]^-$, $[OAc(HOAc)_2]^-$, $[(HSO_4)(H_2SO_4)_2]^-$), amphoteric ($[HSO_4]^-$), as well as basic ($[CH_3COO]^-$, Cl^-, $[Et_2PO_4]^-$, $[N(CN)_2]^-$) properties were selected [17]. The results of PTA solubility as a function of temperature are presented in Figure 4.

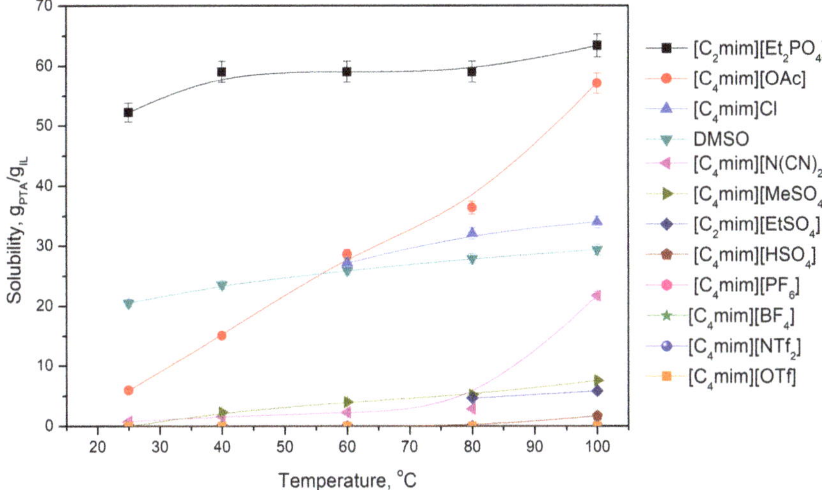

Figure 4. The influence of anion structure on the solubility of PTA.

According to the data presented in this and previous paragraphs, anion structure is crucial in deciding the solubility of PTA. This can be expected, since the chemical properties of ionic liquids are mainly determined by the structure of the anion [20]. The best results were achieved using ionic liquids with diethylphosphate (V) and acetate anions, followed by chloride. Ionic liquids based on these anions possess proton acceptor abilities, which might play a role in the interaction with carboxylic groups from PTA; therefore, enhancing its solubility.

Ionic liquid with the $[N(CN)_2]^-$ anion, which exhibit Lewis base properties, has a moderate ability to dissolve PTA. Nevertheless, the solubility is still better than that for ionic liquids showing weak acidic $[EtSO_3]^-$ or amphoteric $[HSO_4]^-$ properties.

Ionic liquids constructed with acid/base neutral anions—$[BF_4]^-$, $[PF_6]^-$, $[CH_3SO_3]^-$, $[NTf_2]^-$—do not dissolve PTA at all, irrespective of temperature.

2.4. The Solubility of Terephthalic Acid in Protic and Aprotic Ionic Liquids

Aprotic ionic liquids, based on acetate and chloride anions showing good PTA solubility, encouraged us to test their protic analogues, which are cheaper and easier to manufacture. These analogues are formed in a simple reaction between acid and base, for example, acetic acid and 1-methylimidazole. For this purpose, [Hmim][OAc] and [Hmim]Cl ionic liquids were synthesized. Unfortunately, protic ionic liquids show a lower solubility of PTA than their aprotic analogues (Figure 5). It is presumed to be caused by their higher Brønsted acidity, which arises from the presence of a labile proton on the nitrogen atom in the cation.

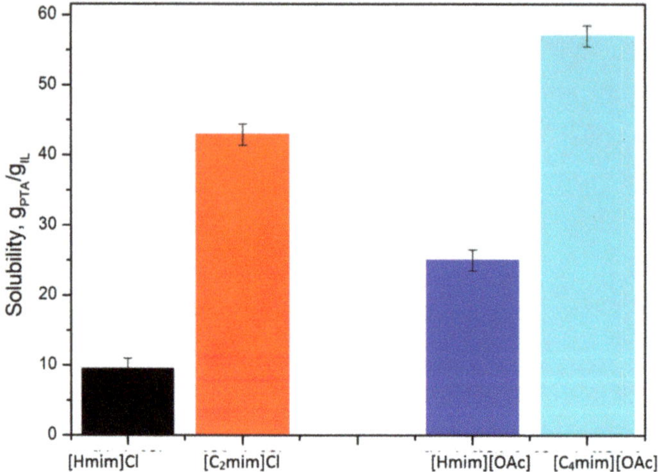

Figure 5. The comparison of the solubility of PTA in protic and aprotic ionic liquids.

3. Materials and Methods

Ionic liquids used in this study: 1-ethyl-3-methylimidazolium diethylphosphate [C$_2$mim][Et$_2$PO$_4$], 1-butyl-3-methylimidazolium acetate [C$_4$mim][OAc], 1-ethyl-3-methylimidazolium chloride [C$_2$mim]Cl, 1-butyl-3-methylimidazolium dicyanamide [C$_4$mim][N(CN)$_2$], 1-butyl-1-methylpyrrolidinium chloride [bmpyr]Cl, tetrabutylammonium bromide [N$_{4444}$]Br, tetradecyl(trihexyl)phosphonium chloride [P$_{14666}$]Cl, tetrabutylammonium chloride [N$_{4444}$]Cl, 1-butyl-3-methylimidazolium methyl sulfate [C$_4$mim][MeSO$_4$], 1-ethyl-3-methylimidazolium ethyl sulfate [C$_2$mim][EtSO$_4$], tetradecyl(trihexyl)phosphonium bromide [P$_{14666}$]Br, 1-butyl-3-methylimidazolium hydrogen sulfate [C$_4$mim][HSO$_4$], 1-butyl-3-methylimidazolium trifluoromethanesulfonate [C$_4$mim][OTf], and 1-butyl-3-methylimidazolium hexafluorophosphate [C$_4$mim][PF$_6$] were purchased from Sigma-Aldrich and dried before use on the Schlenk line (40 °C, 0.5 mbar, 12 h). Other ionic liquids: 1-butyl-3-methylimidazolium chloride [C$_4$mim]Cl, 1-hexyl-3-methylimidazolium chloride [C$_6$mim]Cl, 1-octyl-3-methylimidazolium chloride [C$_8$mim]Cl, 1-methylimidazolium acetate [Hmim][OAc], 1-methylimidazinium chloride [Hmim]Cl, 1-butyl-3-methylimidazolium bis(trifluoromethylsulfonyl)imide [C$_4$mim][NTf$_2$], 1-butyl-3-methylimidazolium tetrafluoroborate [C$_4$mim][BF$_4$], and 1-methylimidazolium hydrogen sulfate [Hmim][HSO$_4$] were synthetized according to the known procedures [7]. PTA was purchased from Sigma-Aldrich and used without further purification.

Solubility measurements: A total of 1 g of ionic liquid was placed in the round bottom flask and then PTA was added in small portions (0.01 g). The mixture was closed under argon atmosphere and stirred using a thermostatic magnetic stirrer for 1 h. When the mixture became homogeneous, the next portion of PTA (0.01 g) was added; if not, the temperature was raised. Measurements were carried out sequentially at 25, 40, 60, 80, and 100 °C.

4. Conclusions

This study systematically expands on the limited existing knowledge concerning the solubility of PTA in ionic liquids. It was confirmed that ionic liquids exhibit a high capacity to dissolve PTA, almost twice higher than the best conventional solvent—DMSO. Ionic liquids based on diethylphosphate, acetate, and chloride anions with Lewis base properties were the most effective. Solubility in ionic liquids decreases in the following order: [C$_2$mim][Et$_2$PO$_4$] > [C$_4$mim][OAc] > [C$_2$mim]Cl > [C$_4$mim]Cl > [C$_6$mim]Cl. Additionally, it was observed that ionic liquids containing an aromatic cation, preferably dialkylimidazolium, performed better than other studied cations. The longer the alkyl substituent in the 1-alkyl-3-methylimidazolium cation, the lower the charge density and polarity, which impedes the solubility of PTA. Additionally, solubility of PTA in ionic liquids strongly depends on the anion structure; the most effective ones possessing an anion with Lewis base properties. In summary, ionic liquids have a high potential to dissolve PTA and can; therefore, be an effective alternative for conventional solvents.

Author Contributions: Conceptualization, K.M., E.P., A.G. and A.C.; Data curation, K.M. and P.L.; Formal analysis, K.M. and P.L.; Funding acquisition, E.P., A.G. and A.C.; Investigation, P.L.; Methodology, K.M., A.G. and A.C.; Project administration, K.M.; Resources, E.P., A.G. and A.C.; Supervision, E.P., A.G. and A.C.; Validation, A.G. and P.L.; Writing—original draft, K.M.; Writing—review and editing, A.C. All authors have read and agreed to the published version of the manuscript.

Funding: This research was funded by Grupa Azoty Zakłady Azotowe Kędzierzyn, S.A.

Conflicts of Interest: The authors declare no conflicts of interest.

References

1. Sheehan, R.J. *Ullmann's Enc. of Industrial Chemistry*; Wiley-VCH: Weinheim, Germany, 2007.
2. Wang, C.L.; Liu, C.L.; Pei, L.X.; Pang, Y.J.; Zhang, Y.; Hou, H.B. Experimental and Modeling Study of Pure Therephthalic Acid (PTA) Wastewater Transport in the Vadose Zone. *Environ. Sci. Impacts.* **2015**, *17*, 389–397. [CrossRef] [PubMed]
3. Acree, W.E.J. IUPAC-NIST Solubility Data Series. 98. Solubility of Polycyclic Aromatic Hydrocarbons in Pure and Organic Solvent Mixtures: Revised and Updated. Part 2. Ternary Solvent Mixtures. *Phys. Chem. Ref. Data.* **2013**, *42*, 1–525.
4. Wasserscheid, P.; Welton, T. *Ionic Liquids in Synthesis*; John Wiley & Sons: Hoboken, NJ, USA, 2007.
5. Jess, A.; Wasserscheid, P. *Chemical Technology*; John Wiley & Sons: Hoboken, NJ, USA, 2012.
6. Weissermel, K.; Arpe, H. *Industrial Organic Chemistry*; John Wiley & Sons: Hoboken, NJ, USA, 2003.
7. Kar, M.; Matuszek, K.; MacFarlane, D.R. *Ionic Liquids in Kirk-Othmer Encyclopedia of Chemical Technology*; John Wiley & Sons: Hoboken, NJ, USA, 2019.
8. Welton, T. Ionic liquids: A brief history. *Biophys. Rev.* **2018**, *10*, 691–706. [CrossRef] [PubMed]
9. Procopio, A.; Costanzo, P.; Curini, M.; Nardi, M.; Oliverio, M.; Sindona, G. Erbium(III) Chloride in Ethyl Lactate as a Smart Ecofriendly System for Efficient and Rapid Stereoselective Synthesis of *trans*-4,5-Diaminocyclopent-2-enones. *ACS Sustain. Chem. Eng.* **2013**, *1*, 541–544. [CrossRef]
10. Nardi, M.; Costanzo, P.; De Nino, A.; Di Gioia, M.L.; Olivito, F.; Sindona, G.; Procopio, A. Water excellent solvent for the synthesis of bifunctionalized cyclopentanones from furfural. *Green Chem.* **2017**, *19*, 5403–5411. [CrossRef]
11. Ahmed, M.I.; Asiri, A.M. *Industrial Applications of Green Solvents–Volume I*; Inamuddin, M.I.A., Asiri, A.M., Eds.; Materials Research Forum LLC: Millersville, PA, USA, 2019.
12. Di Gioia, M.L.; Costanzo, P.; De Nino, A.; Maiuolo, L.; Nardi, M.; Olivito, F.; Procopio, A. Simple and efficient Fmoc removal in ionic liquid. *RSC Adv.* **2017**, *7*, 36482–36491. [CrossRef]
13. Geldbach, T.J.; Dyson, P.J. *Metal Catalysed Reactions in Ionic Liquids*; Springer: Heidelberg, Germany, 2005.
14. Rogers, R.D.; Myerson, A.S.; Corey Hines, C. Process for Purification of Aryl Carboxylic Acids US 2010174111 A1, 8 July 2010.
15. Martins, S.C.; DeSalvo, K.; Bhattacharyya, A. Process for Purifying Terephthalic Acid Using Ionic Liquids US 2013331603 A1, 12 December 2013.
16. Aduri, P.; Uppara, P.V.; Jain, S.S. Process for Separating Aryl Carboxylic Acids US 20150065748 A1, 4 March 2015.
17. Brown, I.D. Structural chemistry and solvent properties of dimethyls. *J. Solution Chem.* **1987**, *16*, 205–224. [CrossRef]
18. Matthews, R.P.; Welton, T.; Hunt, P.A. Hydrogen bonding and π-π interactions in imidazolium-chloride ionic liquid clusters. *Phys. Chem. Chem. Phys.* **2015**, *17*, 14437–14453. [CrossRef] [PubMed]

19. Zhou, T.; Chen, L.; Ye, Y.; Chen, L.; Qi, Z.; Freund, H.; Sundmacher, K. An overview of mutual solubility of ionic liquids and water predicted by COSMO-RS. *Ind. Eng. Chem. Res.* **2012**, *51*, 6256–6264. [CrossRef]

20. MacFarlane, D.R.; Pringle, J.M.; Johansson, K.M.; Forsyth, S.A.; Forsyth, M. Lewis base ionic liquids. *Chem. Commun.* **2006**, 1905–1917. [CrossRef] [PubMed]

Sample Availability: Samples of ionic liquids are available from the authors.

 *molecules* 

Article

Sustainable and Selective Extraction of Lipids and Bioactive Compounds from Microalgae

Ilaria Santoro [1,*], Monica Nardi [2,*] , Cinzia Benincasa [3] , Paola Costanzo [2] , Girolamo Giordano [1], Antonio Procopio [2] and Giovanni Sindona [4]

1   Dipartimento di Ingegneria per l'Ambiente e il Territorio e Ingegneria Chimica, Università della Calabria, Cubo 45A, I-87036 Rende, Italy; girolamo.giordano@unical.it
2   Dipartimento di Scienze della Salute, Università Magna Græcia, Viale Europa, I-88100 Germaneto (CZ), Italy; pcostanzo@unicz.it (P.C.); procopio@unicz.it (A.P.)
3   CREA Research Centre for Olive, Citrus and Tree Fruit, C.da Li Rocchi, I-87036 Rende, Italy; cinzia.benincasa@crea.gov.it
4   Dipartimento di Chimica e Tecnologie Chimiche, Università della Calabria, Cubo 12C, I-87036 Rende, Italy; giovanni.sindona@unical.it
*   Correspondence: ilaria.santoro@unical.it (I.S.); monica.nardi@unicz.it (M.N.); Tel.: +39-0961-3694116 (M.N.)

Academic Editors: Monica Nardi, Antonio Procopio and Maria Luisa Di Gioia
Received: 7 November 2019; Accepted: 27 November 2019; Published: 28 November 2019

Abstract: The procedures for the extraction and separation of lipids and nutraceutics from microalgae using classic solvents have been frequently used over the years. However, these production methods usually require expensive and toxic solvents. Based on our studies involving the use of eco-sustainable methodologies and alternative solvents, we selected ethanol (EtOH) and cyclopentyl methyl ether (CPME) for extracting bio-oil and lipids from algae. Different percentages of EtOH in CPME favor the production of an oil rich in saturated fatty acids (SFA), useful to biofuel production or rich in bioactive compounds. The proposed method for obtaining an extract rich in saturated or unsaturated fatty acids from dry algal biomass is disclosed as eco-friendly and allows a good extraction yield. The method is compared both in extracted oil percentage yield and in extracted fatty acids selectivity to extraction by supercritical carbon dioxide (SC-CO_2).

Keywords: algal oil; green chemistry; green solvents; extraction; biofuel; bio compound

1. Introduction

In recent years, the production of algae culture and usage of algal biomass conversion products have received much attention. Microalgae are a potential source of a wide range of high-value products for different biotechnological uses [1]. In particular, algae have long been considered excellent feedstock to produce oils. Algal oil, in fact, can be used in different sectors in addition to the production of biofuels [2], for example in the nutraceutical sector as nutritional supplements and in cosmetics.

The considerable amounts of lipid content in microalgae allow the production of alternative renewable cleaner fuels [3].

Biodiesel is a mixture of fatty acid alkyl esters usually obtained by transesterification (ester exchange reaction) of vegetable oils or animal fats [4]. Many research reports and articles have described different advantages of using microalgae for biodiesel production in comparison with other available feedstocks [5–12]. The lipid and fatty acid substances of microalgae differ in accordance with culture conditions. In fact, depending on the strain to which the algae belong, it can have between 20–80% of oil by weight of dry mass [13], and it also varies in the lipid composition [14].

From a practical point of view, microalgae are easy to cultivate, can grow with little or even no attention, use water unsuitable for human consumption, and are easily inclined to provide nutrients.

The extracts of microalgae show antimicrobial, antiviral, antibacterial, and antifungal properties attributed to the presence of fatty acids [15] and are also used as ingredients in different skincare, sun protection, and hair care formulations. Microalgae are considered, in fact, as the predominant production sources for polyunsaturated fatty acids (PUFAs) that have to be supplied for the human diet [16,17]. PUFAs have been used in the prevention/treatment of cardiovascular diseases [18–21] and their derivatives, namely α-linolenic acid (ALA), eicosapentaenoic acid (EPA), docosapentaenoic acid (DPA), and docosahexaenoic acid (DHA), have also been reported for the treatment of type 2 diabetes, inflammatory bowel disorders, skin disorders, and asthma [22–24].

PUFAs play a major role in the treatment of arthritis, obesity, Parkinson's disease, and heart disease [25]. EPA and DHA are the main derivatives of omega-3 fatty acids (PUFA n-3) and play a role in lowering blood cholesterol and in fetal brain development, respectively [26]. Carotenoids and pigments are the main constituents of microalgal-based food supplements, and they possess antioxidant activities with neuroprotective action and protection against chronic diseases [27].

Various extraction methods have been reported in the literature for micro-algal lipids. Conventional methods for extraction lipids include hexane extraction and vacuum distillation. The traditional solvent extraction is the most used method thanks to the simplicity in operation and the possibility of use in the industrial field [28]. Usually, solvents such as methanol and chloroform, and temperatures between 150 °C and 250 °C, are used to obtain high extractive yields of microalgae oil [29,30]. The Bligh and Dyer method is the one most used in the extraction and quantitation of lipids at the analytical level [29]. The use of flammable or toxic solvents is considered a very important problem due to the adverse health and environmental effects.

Over the years, several research groups have determined the profile of triglycerides in the oil extracted from microalgae, which is relevant for the production of biofuels [9,30–33]. New algae oil extraction techniques are being developed, such as enzyme-assisted extraction [34], microwave-assisted extraction [35], ultrasound-assisted extraction [36], pressurized liquid extraction [37], and supercritical fluid extraction [38,39]. Recent studies have shown that total lipids/bio-oil extraction from algae can occur through using supercritical carbon dioxide (SC-CO_2) assisted with azeotropic co-solvents such as hexane and ethanol 1/1 at a reaction pressure of 340 bar, and a temperature of 80 °C in 60 min, obtaining a total algal lipid yield of 31.37% based on dry basis and a percentage of eicosapentanoic acid (EPA) in the range of 20% to 32% [40]. This procedure increased the total lipid yield and the selectivity, but the usage of hexane as a solvent lead to numerous consequences such as air pollution and toxicity.

Our research group has done extensive work on the identification and molecular characterization of food compounds [41–43]. We have developed environmentally friendly methods for the extraction and further chemical manipulation of natural bioactive molecules [44–52], reducing or eliminating the use and generation of harmful substances and solvents aiming to encourage green chemistry [53]. In this study, according to the previous studies based on the use of non-toxic solvents [54–64], we have focused our attention on the development of selective extractive processes for PUFAs rather than for saturated fatty acids and vice versa, using green solvents such as Cyclopentyl Methyl Ether (CPME) and ethanol (EtOH).

CPME is an unconventional and an ethereal green solvent, which is very stable to peroxide formation, with low volatility and low water solubility [65]. Thus, it has increased interest as an industrial solvent [66]. It exerts no genotoxity or mutagenicity [66], and is produced by the addition of methanol to cyclopentene, with a 100% atom economy for its synthesis. It has also been studied in many important chemical processes including furfural synthesis [67] and extraction of natural products [68]. Previous works studied the liquid–liquid equilibria for ternary systems of water/CPME/alcohol (methanol, ethanol, 1-propanol, or 2-propanol) to test greener solvent systems that substitute and simulate the Bligh and Dyer method for oil extraction [69]. Recently, a comparative study of lipid extraction from wet microalgae was performed using several methods such as the Soxhlet, Bligh and Dyer, Folch, and Hara and Radin methods, with 2-methyltetrahydrofuran (2-MeTHF) and CPME as green solvents. The Bligh and Dyer methodology using the solvents 2-MeTHF/isoamyl alcohol (2:1 *v/v*) and CPME/methanol (1:1.7 *v/v*) showed an oil extraction yield less than 10% [70].

Thus, the objective of this work was to evaluate the use of a green binary solvent system CPME/EtOH for the selective extraction of lipids and bioactive compounds from microalgae dry. The various lipid components within each fraction have been characterized and quantified using GC-MS and LC-MS [71–73], following a standard procedure of transmethylation [74] available to our goal. The information on complete lipid characterization of extracts is essential for the successful selection of the extraction process that is useful for the production of biofuels or the development of potential nutraceuticals.

2. Results and Discussion

The biodiesel or saturated fatty acids production process involves different steps, including lipid extraction and purification of fatty acids. The classic extraction processes often involve the use of toxic substances. Furthermore, the selective extraction processes that leads to obtaining oils rich in PUFAs or saturated fatty acids, and the separation of individual fatty acids, are difficult for the production of highly concentrated w-3 components. Docosahexaenoic acid (DHA, 22:6w3) is considered to be a crucial nutrient for fetal and infant development [75,76], and only recently researchers have developed the urea complexation to concentrate DHA from Crypthecodinium cohnii CCMP 316 biomass [77] and the production of algal oil enriched in w-3-PUFA [78,79] that is useful for human health.

The use of SC-CO$_2$ is an environmentally sustainable extraction method [39]. However, to have a higher extraction yield, it is necessary to use co-solvents such as hexane [40], invalidating the sustainability of the method itself. Moreover, this method is not always easily applicable at an industrial level.

In the present work, according to the studies based on the use of non-toxic solvent [80,81], extracted lipids rich in PUFAs or saturated fatty acids by dried microalgae were obtained using CPME and ethanol as green solvents at different percentages. The optimization of the method was carried out by comparing the oil extraction yield obtained using different percentages of ethanol in CPME with the oil extraction yield obtained by SC-CO$_2$ [38]. As can be shown in Figure 1, the extraction yield increased considerably with the use of 80% of ethanol in CPME (39.4%).

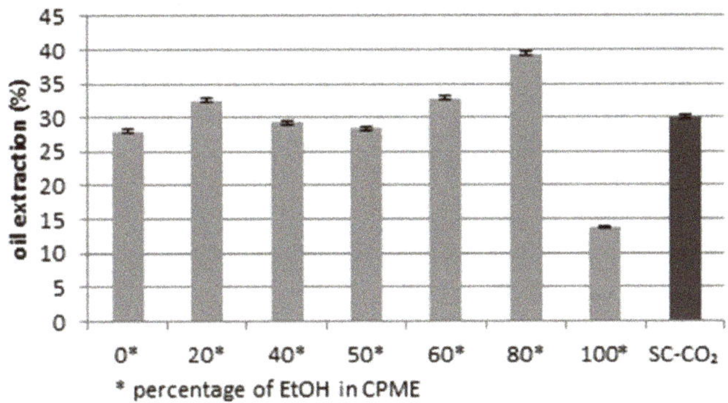

Figure 1. Oil extraction yields (g for 100 g of dry matter) using different volume percentages of ethanol in cyclopentyl methyl ether (CPME), compared to the yield obtained by SC-CO$_2$. $p < 0.05$.

The determination of fatty acid composition using GC-MS quantitative analysis in the extracted samples (see supporting information) was performed. The GC-MS analysis of the methyl transesterified [82] algal oil was extracted using SC-CO$_2$ (see supporting information, S1), revealing the presence of fourteen fatty acids (Table 1). At the time retention of 13.35 min, the peak of internal standard (methyl tricosanoate, 23:0) was observed.

Peak identification of the saturated fatty acids (SFA) and unsaturated fatty acids (PUFAs and MUFA) in the analyzed microalgae oil samples were carried out by comparison with retention time and mass spectra of known standards (Table 1). Samples were analyzed in triplicate.

Table 1. Fatty acids composition of algal oil by GC-MS analysis.

S/N	RT (min)	Name of Compound	Mol. Formula	Classification
1	1.5	Myristic acid (14:0)	$C_{14}H_{28}O_2$	SFA
2	1.7	Pentadecylic acid (15:0)	$C_{15}H_{30}O_2$	SFA
3	2.1	Palmitic acid (16:0)	$C_{16}H_{32}O_2$	SFA
4	2.7	Heptadecanoic acid (17:0)	$C_{17}H_{34}O_2$	SFA
5	3.1	Linoleic acid (18:2 ω-6)	$C_{18}H_{32}O_2$	PUFA
6	3.2	γ-Linolenic acid (18:3 ω-6)	$C_{18}H_{30}O_2$	PUFA
7	3.4	Oleic acid (18:1 ω-9)	$C_{18}H_{34}O_2$	MUFA
8	3.5	Stearic acid (18:0)	$C_{18}H_{36}O_2$	SFA
9	5.4	Eicosapentaenoic acid, EPA (20:5 ω-3)	$C_{20}H_{30}O_2$	PUFA
10	5.5	Eicosatrienoic acid (20:3 ω-6)	$C_{20}H_{34}O_2$	PUFA
11	5.6	Eicosatetraenoic acid (20:4 ω-6)	$C_{20}H_{32}O_2$	PUFA
12	6.2	Eicosanoic acid (20:0)	$C_{20}H_{40}O_2$	PSFA
13	9.1	Docosapentaenoic acid, DPA, (22:5 ω-3)	$C_{22}H_{34}O_2$	PUFA
14	9.4	Docohexaenoic acid, DHA, (22:6 ω-3).	$C_{22}H_{32}O_2$	PUFA

From the GC-MS analysis, it was found that the most abundant fatty acid methyl esters (FAMEs) present in the various extracts were those of myristic and palmitic acid (two saturated fatty acids), and EPA and DHA (two ω-3 polyunsaturated fatty acids) derivatives. These were considered to evaluate the performance of the EtOH/CPME mixture compared to SC-CO$_2$ in the extraction process (Figure 2).

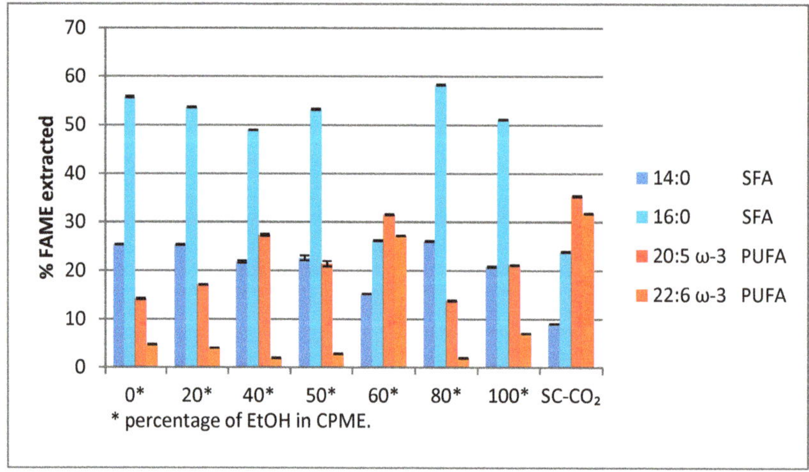

Figure 2. % of myristic acid, palmitic acid (saturated fatty acids, SFAs), eicosapentaenoic acid (EPA), and docosahexaenoic acid (DHA) (polyunsaturated fatty acids, PUFAs) extracted with a mixture ethanol (EtOH)/CPME and SC-CO$_2$. $p < 0.05$.

It is evident that an oil richer in saturated fatty acids (Figure 2) has been obtained using a mixture of EtOH/CPME 8/2, the same solution solvent useful for obtaining a higher extraction yield of algal oil.

On the contrary, an oil rich in EPA and DHA was obtained using a mixture of EtOH/CPME 6/4 (Figure 2), the same mixture of solvents, which showed a yield of the extracted oil equal to 32.8% against 30.0% yield obtained with SC-CO$_2$.

From the results of the quantitative analysis of all fatty acid methyl esters (see Supplementary Material, S2), the yield percentage of all saturated fatty acids and all unsaturated fatty acids was calculated.

Table 2 illustrates how the percentage of total SFA and total UFA (polyunsaturated and monounsaturated fatty acids) concentrations varied according to the percentage of EtOH in CPME (Table 2, entries 1–7) and using SC-CO$_2$ (Table 2, entry 8). The use of SC-CO$_2$ favored the isolation of UFA to SFA (66.93%). A comparable result was obtained using an EtOH/CPME 6/4 mixture (61.41%) useful for obtaining a higher oil extraction yield.

Table 2. SFA (saturated fatty acids) and UFA (unsaturated fatty acids) percentage concentration variation as a function of solvent/solvent mixture used.

Entry	Extraction Method	% SFA	% UFA
1	0 *	78.25 ± 0.16	21.75± 0.16
2	20 *	73.34 ± 0.04	26.66± 0.06
3	40 *	65.34 ± 0.09	34.66± 0.07
4	50 *	69.80 ± 0.15	30.20± 0.10
5	60 *	38.59 ± 0.03	61.41± 0.05
6	80 *	82.83 ± 0.13	17.17± 0.09
7	100 *	71.46 ± 0.01	28.54± 0.07
8	SC-CO$_2$ **	33.06 ± 0.18	66.93± 0.09

* Percentage of EtOH in CPME. Solvent mixture used for Soxhlet system. ** Supercritical carbon-dioxide extraction method. The values of percentages are in mean ± SD (n = 3).

Electron ionization (EI) MS coupled to Gas chromatography (GC) to analyze fatty acids [83] has been usually applied but in addition to GC-MS, liquid chromatography (LC)-MS was also a useful method for the accurate analysis of profiling of FFAs [84–86] and for qualitative determination of nonvolatile compounds in food [87]. The use of electrospray ionization (ESI) MS, a soft ionization technique, provided the information on molecular ions. Therefore, tandem MS (MS/MS) was applied for the most sensitive and selective analysis of FFAs.

To provide the identification of components in the extracted samples, liquid-chromatography mass spectrometry (LC-MS) was employed. ESI-MS/MS analysis was performed for all the ions present in the full scan chromatogram for each algae extract.

The analytical technique thus confirmed in a more detailed way the presence of the previous extracted and previously quantified fatty acids. The chromatogram was obtained scanning between 50 and 800 amu in negative ion mode (Figure 3).

From the information obtained in full scan mode, it was possible to have information about the molecular weights of the fatty acids possibly present in the samples under investigation.

From the molecular ions registered, the algal oil extract was composed of lauric acid (molecular ion at *m/z* 199.4); myristic acid (227.4); myristoleic acid (225.2); pentadecylic acid (241.4); palmitic acid (255.5); γ-linolenic acid (277.4); oleic acid (281.6); stearic acid (283.5); eicosapentanoic acid (301.6); docosaexaenoic acid (327.5); docosapentaenoic acid (329.7) and behenic acid (339.6). Successively, to attribute the molecular structures on each single molecular ion previously recorded, experiments in product ion scan (MS/MS) were performed.

Table 3 lists the deprotonated molecules identified in full scan MS spectra, the fragment ions, and the precursor ions identified by MS/MS experiments.

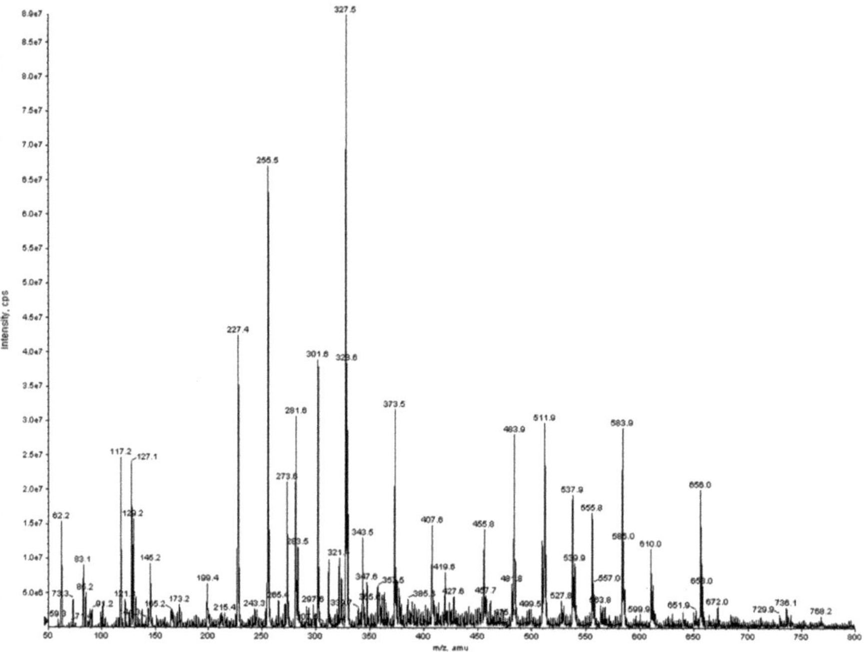

Figure 3. Negative electrospray ionization (ESI) full scan mass spectrum of algal oil sample.

Table 3. List of the deprotonated molecules identified in the full scan MS spectra of algal oil sample, fragment ions, and precursor ions identified in MS/MS spectra.

Analyte	[M − H]⁻	PIS	PREC
Lauric acid	199.4	181.6; 155.0	399.7; 455.9
Myristic acid	227.4	209.5; 183.3	455.7; 483.8; 509.8
Myristoleic acid	225.2	207.2; 181.5	482
Pentadecylic acid	241.4	223.2; 197.2	483.8
Palmitic acid	255.5	237.6; 211.6	511.8; 537.9; 583.9
γ-linolenic acid	277.4	259.4; 233.4	555.8
Oleic acid	281.6	263.4; 237.4	509.8; 537.9; 563.8
Stearic acid	283.5	265.4; 239.2	540.5; 568
Eicosapentanoic acid, EPA	301.6	283.6; 257.6	602.2
Docosahexaenoic acid, DHA	327.5	283.6; 309.2; 229.6	583.9; 610.0; 656.0; 658.0
Docosapentaenoic acid, DPA	329.7	311.7; 285.6	658
Behenic acid	339.6	321.3; 294.9	596.5; 569.2

The initial complexity of the mass spectrum was, therefore, reduced when the product ion scan and a precursor ion scan were performed. ESI-MS/MS analysis was carried out for all the ions present in the full scan chromatogram for each algae extract. Ions included in the range of *m/z* 455–656 indicated molecular adducts between two identical or different fatty acid compounds (see Supplementary Material, Table S3).

Based on this result, it was possible to choose the type of solvent/solvent mixture to be used depending on the type of extract to be obtained (richer than SFA or UFA). Both analytical techniques were useful and necessary to determine the type of fatty acids present in the extracted samples. Thus, it was possible to identify an environmentally sustainable extraction method for saturated or unsaturated fatty acids with excellent extraction yield and significant selectivity.

3. Materials and Methods

3.1. Chemicals

Solvents, reagents, and thimbles were purchased from Sigma–Aldrich (Sigma–Aldrich, St. Louis, MO, USA). Dry algal biomass was provided by Aquafauna Biomarine inc. (P.O. Box 5, Hawthorne, CA, USA).

3.2. Supercritical Fluid Extraction

Fractionation of algal oil from algal biomass was carried out in a continuous extraction process supercritical fluid using an Applied Separations Speed SFE model 7070 apparatus (Applied Separations Inc., Allentown, PA, USA). About 5 g of sample and glass beads (size about 3 mm) were weighed and added to the vessel (1.27 cm i.d. × 25.4 cm long) and sealed with polypropylene wool at the top and the bottom of the extraction vessel. The oven temperature used was 80 °C and the vessel at a pressure of 35 MPa and room temperature of 25 °C. The extracted oils were collected within 10 mL vials. Each extraction was replicated 3 times.

3.3. Soxhlet Extraction

Soxhlet Extraction was carried out using 100 mL of solvent on 5 g of dried algae sample by Soxhlet apparatus. The extraction lasted for 1 h. The extract was filtered to remove possible solid particles. Organic solutions were then concentrated by rotary evaporation, and the traces of solvent in residual oil were removed by nitrogen flushing. The yield was calculated based on the weight of extracted oil and the weight of the start sample. The same process was repeated with different solvent mixtures; CPME 100%, CPME/EtOH (80:20, 60:40, 50:50, 40:60, 20:80) and EtOH 100%. Each extraction was replicated 3 times.

3.4. Preparation of Methyl Esters of Constituent Fatty Acids

Fatty acids compositions were determined by their conversion to methyl esters. 15 mg of each oil was added to the internal standard (250 ng/100 µL chloroform, methyl tricosanoate, 23:0). The oil was subjected to transmethylation by treating 15 mg of each oil with 6 mL of 0.2 M sulfuric acid in methanol following standard procedure and 15 mg of hydroquinone [64]. The mixture was incubated for 12 h at 60 °C and subsequently cooled. 1 mL of distilled water was added to each vial and extracted 3 times with 1.5 mL of heptane. The organic phase containing the fatty acid methyl esters (FAMEs) was separated and evaporated under a stream of nitrogen. The FAMEs obtained were dried over anhydrous magnesium sulfate and kept for Gas Chromatography-Mass Spectrometry (GC-MS, Shimadzu, Kyoto, Japan) analysis.

3.5. GC-MS Analysis

The FAMEs obtained from the different algal oils were analyzed on Shimadzu GC/MS-QP 2010 gas chromatography instrument with an autosampler AOC-20i (Shimadzu) equipped with a 30 m-QUADREX 007-5MS capillary column operating in the "split" mode, and 1 mL min-1 flow of He as carrier gas. The injector was operated at 250 °C, while the detector was operated at 380 °C. The oven temperature was programmed to rise from 70 to 135 °C at a heating rate of 2 °C/min, from 135 to 220 °C at a heating rate of 4 °C/min and from 220 to 270 °C at a heating rate of 3.5 °C/min. The mass spectrometer (Shimadzu, Kyoto, Japan) was operated in the electron impact (EI) mode at 70 eV in the scan range 50–500 m/z. The FAMEs were identified based on the authentic samples previously injected in combination with the examination with individual molecular weight, mass spectra, and comparison of fragmentation pattern in the mass spectrum with that of the National Institute of Science and Technology, NIST library.

3.6. Data Analysis

Results were analyzed by using a one-way ANOVA (GraphPad Software Inc., San Diego, CA, USA), and data were presented as the mean ± standard error of mean (SEM), except otherwise indicated. All experiments were performed in triplicate. Values at $p < 0.05$ were taken as significant.

3.7. Flow Injection Analysis/Mass Spectrometry (FIA/MS)

The mass spectrometer system used for the qualitative analyses of the algal oil extracts was a Q-Trap API 4000 (MSD Sciex Applied BiosystemFoster City, CA). The analyses were performed by flow injection analyses (FIA) in both negative and positive ion modes under a flow rate of 10 μL/min. In general, the intensity of ions observed as positive ions were lower than those observed as negative ions. Therefore, the experiments were conducted using the negative mode. The most important point of this experiment was to confirm that ions generated from the target compounds were observed, rather than that merely any sort of ions were observed. Therefore, it was necessary to attribute observed ions to specific compounds. Fundamentally, in the attribution process, assuming that positive ions were protonated molecules $[M + H]^+$ and negative ions were deprotonated molecules $[M - H]^-$, it was verified whether they were consistent with the molecular mass of target compounds. The structural assignment was, therefore, based on the accurate mass of the pseudo-molecular ion $[M - H]^-$, present in the negative ESI-MS chromatogram, and on the corresponding fragment ions detected by collision-induced dissociation (CID) under nitrogen (25% normalized collision energy) in the ion trap. The instrument parameters were set as follows: Ion spray voltage (IS) −4600 V; curtain gas 10 psi; ion source gas 12 psi; collision gas thickness medium; entrance potential 10 eV, declustering potential 70 eV, collision energy (CE) between 15 and 30 eV and collision exit potential (CXP) between 5 and 9 eV.

4. Conclusions

In this work, a new eco-friendly and effective method for extracting total lipids/bio-oil from algae was shown. The maximum total algal lipid yield (39.4% based on dry basis) was obtained using a solution of EtOH/CPME (8:2) at a temperature of 80 °C and a reaction time of 60 min. This extraction condition is advantageous to obtain an oil rich in SFA, useful to biofuel production. An oily extract rich in bio-compounds that contribute to human well-being such as eicosapentaenoic acid, EPA (C20: 5, ω-3) and docosahexaenoic acid, DHA (C22: 6, ω-3) was obtained using a EtOH/CPME, 6:4, at the same temperature and reaction time. In these extraction conditions the oil extraction yield is equal to 32.8% against 30.0% if SC-CO$_2$ is used.

This method offers many advantages over conventional extraction technologies, such as increased total lipid yield, increased selectivity, and preserved thermo-labile compounds. It is an alternative method for sustainable production of algal biofuel and the development of high-value co-products.

Supplementary Materials: The following are available online, GC-MS and MS (EI) spectra of products, quantitative analysis of all fatty acid methyl esters, negative ESI full scan mass spectrum of algal oil sample, negative ESI full scan mass spectrum of standard solution of DHA, Stearic acid and butyric acid, negative ESI full scan mass spectrum of standard solution of oleic alcohol and palmitic acid.

Author Contributions: Conceptualization, M.N. and I.S.; formal analysis, C.B. and P.C.; project administration, G.S. and G.G.; validation, G.S. and G.G.; Supervision, A.P.

Funding: Work was supported by Dipartimento di Scienze della Salute, Università Magna Græcia, Italy, and by the Calabria region under the project APQ-RAC QUASIORA. Ilaria Santoro thanks the doctorate school of Translational Medicine for a fellowship.

Conflicts of Interest: The authors declare no conflict of interest.

References

1. Chu, W.L. Biotechnological applications of microalgae. *IeJSME* **2012**, *6*, S24–S37.
2. Schenk, P.M.; Thomas-Hall, S.R.; Stephens, E.; Marx, U.C.; Mussgnug, J.H.; Posten, C.; Hankamer, B. Second generation biofuels: High-efficiency microalgae for biodiesel production. *Bioenergy Res.* **2008**, *1*, 20–43. [CrossRef]
3. Hossain, A.S.; Salleh, A.; Boyce, A.N.; Chowdhury, P.; Naqiuddin, M. Biodiesel fuel production from algae as renewable energy. *Am. J. Biochem. Biotechnol.* **2008**, *4*, 250–254. [CrossRef]
4. Medina, A.R.; Cerdán, L.E.; Giménez, A.G.; Páez, B.C.; González, M.I.; & Grima, E.M. Lipase-catalyzed esterification of glycerol and polyunsaturated fatty acids from fish and microalgae oils. *J. Biotechnol.* **1999**, *70*, 379–391. [CrossRef]
5. Li, Y.; Horsman, M.; Wu, N.; Lan, C.Q.; Dubois-Calero, N. Biofuels from microalgae. *Biotechnol. Prog.* **2008**, *24*, 815–820. [CrossRef] [PubMed]
6. Li, Y.; Horsman, M.; Wang, B.; Wu, N.; Lan, C.Q. Effects of nitrogen sources on cell growth and lipid accumulation of green alga Neochloris oleoabundans. *Appl. Microbiol. Biotechnol.* **2008**, *81*, 629–636. [CrossRef] [PubMed]
7. Sheehan, J.; Dunahay, T.; Benemann, J.; Roessler, P. *Look Back at the US Department of Energy's Aquatic Species Program: Biodiesel from Algae*; close-out report (No. NREL/TP-580-24190); National Renewable Energy Lab: Golden, CO, USA, 1998.
8. Chisti, Y. Biodiesel from microalgae. *Biotechnol. Adv.* **2007**, *25*, 294–306. [CrossRef]
9. Hu, Q.; Sommerfeld, M.; Jarvis, E.; Ghirardi, M.; Posewitz, M.; Seibert, M.; Darzins, A. Microalgal triacylglycerols as feedstocks for biofuel production: Perspectives and advances. *Plant J.* **2008**, *54*, 621–639. [CrossRef]
10. Rodolfi, L.; Chini Zittelli, G.; Bassi, N.; Padovani, G.; Biondi, N.; Bonini, G.; Tredici, M.R. Microalgae for oil: Strain selection, induction of lipid synthesis and outdoor mass cultivation in a low-cost photobioreactor. *Biotechnol. Bioeng.* **2009**, *102*, 100–112. [CrossRef]
11. Rosenberg, J.N.; Oyler, G.A.; Wilkinson, L.; Betenbaugh, M.J. A green light for engineered algae: Redirecting metabolism to fuel a biotechnology revolution. *Curr. Opin. Biotechnol.* **2008**, *19*, 430–436. [CrossRef]
12. Tsukahara, K.; Sawayama, S. Liquid fuel production using microalgae. *J. Jpn. Pet. Inst.* **2005**, *48*, 251–259. [CrossRef]
13. Bajhaiya, A.K.; Mandotra, S.K.; Suseela, M.R.; Toppo, K.; Ranade, S. Algal Biodiesel: The next generation biofuel for India. *Asian J. Exp. Biol. Sci.* **2010**, *1*, 728–739.
14. Pratoomyot, J.; Srivilas, P.; Noiraksar, T. Fatty acids composition of 10 microalgal species. *Songklanakarin J. Sci. Technol.* **2005**, *27*, 1179–1187.
15. Kamenarska, Z.; Serkedjieva, J.; Najdenski, H.; Stefanov, K.; Tsvetkova, I.; Dimitrova-Konaklieva, S.; Popov, S.; Kamenarska, Z.; Serkedjieva, J.; Najdenski, H.; et al. Antibacterial, antiviral, and cytotoxic activities of some red and brown seaweeds from the Black Sea. *Bot. Mar.* **2009**, *52*, 80–86. [CrossRef]
16. Ward, O.P.; Singh, A. Omega-3/6 fatty acids: Alternative sources of production. *Process Biochem* **2005**, *40*, 3627–3652. [CrossRef]
17. Handayani, N.A.; Ariyanti, D.; Hadiyanto, H. Potential Production of Polyunsaturated Fatty Acids from Microalgae. *Int. J. Eng. Sci.* **2011**, *2*, 13–16. [CrossRef]
18. Roynette, C.E.; Calder, P.C.; Dupertuis, Y.M.; Pichard, C. *n*-3 polyunsaturated fatty acids and colon cancer prevention. *Clin. Nutr.* **2004**, *23*, 139–151. [CrossRef]
19. Simopoulos, A.P. The importance of the omega-6/omega-3 fatty acid ratio in cardiovascular disease and other chronic diseases. *Exp. Biol. Med.* **2008**, *233*, 674–688. [CrossRef]
20. Albert, C.M.; Campos, H.; Stampfer, M.J.; Ridker, P.M.; Manson, J.A.; Willett, W.C. Blood Levels of Long-Chain n–3 Fatty Acids and the Risk of Sudden Death. *N. Engl. J. Med.* **2002**, *346*, 1113–1118. [CrossRef]
21. Shahidi, F.; Miraliakbari, H. Omega-3 (*n*-3) fatty acids in health and disease: Part 1-cardiovascular disease and cancer. *J. Med. Food* **2004**, *7*, 387–401. [CrossRef]
22. Hartweg, J.; Perera, R.; Montori, V.M.; Dinneen, S.F.; Neil, A.H.; Farmer, A.J. Omega-3 polyunsaturated fatty acids (PUFA) for type 2 diabetes mellitus. *Cochrane Database Syst. Rev.* **2008**. [CrossRef] [PubMed]

23. Oliver, E.; McGillicuddy, F.; Phillips, C.; Toomey, S.; Roche, H.M. The role of inflammation and macrophage accumulation in the development of obesity-induced type 2 diabetes mellitus and the possible therapeutic effects of long-chain *n*-3 PUFA. *P. Nutr. Soc.* **2010**, *69*, 232–243. [CrossRef] [PubMed]

24. Miller, C.C.; Tang, W.; Ziboh, V.A.; Fletcher, M.P. Dietary supplementation with ethyl ester concentrates of fish oil (*n*-3) and borage oil (*n*-6) polyunsaturated fatty acids induces epidermal generation of local putative anti-inflammatory metabolites. *J. Invest. Dermatol.* **1991**, *96*, 98–103. [CrossRef] [PubMed]

25. Fernando, I.S.; Nah, J.W.; Jeon, Y.J. Potential anti-inflammatory natural products from marine algae. *Environ. Toxicol. Pharmacol.* **2016**, *48*, 22–30. [CrossRef] [PubMed]

26. McCann, J.C.; Ames, B.N. Is docosahexaenoic acid, an n-3 long-chain polyunsaturated fatty acid, required for development of normal brain function? An overview of evidence from cognitive and behavioral tests in humans and animals. *Am. J. Clin. Nutr.* **2005**, *82*, 281–295. [CrossRef] [PubMed]

27. Pangestuti, R.; Kim, S.K. Biological activities and health benefit effects of natural pigments derived from marine algae. *J. Funct. Foods* **2011**, *3*, 255–266. [CrossRef]

28. Topare, N.S.; Raut, S.J.; Renge, V.C.; Khedkar, S.V.; Chavanand, Y.P.; Bhagat, S.L. Extraction of oil from algae by solvent extraction and oil expeller method. *Int. J. Chem. Sci.* **2011**, *9*, 1746–1750.

29. Bligh, E.G.; Dyer, W.J. A rapid method of total lipid extraction and purification. *Can. J. Biochem. Physiol.* **1959**, *37*, 911–917. [CrossRef]

30. McNichol, J.; MacDougall, K.M.; Melanson, J.E.; McGinn, P.J. Suitability of soxhlet extraction to quantify microalgal fatty acids as determined by comparison with in situ transesterification. *Lipids* **2012**, *47*, 195–207. [CrossRef]

31. Wyman, C.E.; Goodman, B.J. Biotechnology for production of fuels, chemicals, and materials from biomass. *Appl. Biochem. Biotech.* **1993**, *39*, 41. [CrossRef]

32. Demirbas, A.; Demirbas, M.F. Importance of algae oil as a source of biodiesel. *Energ. Convers. Manag.* **2011**, *52*, 163–170. [CrossRef]

33. Talebi, A.F.; Mohtashami, S.K.; Tabatabaei, M.; Tohidfar, M.; Bagheri, A.; Zeinalabedini, M.; Bakhtiari, S. Fatty acids profiling: A selective criterion for screening microalgae strains for biodiesel production. *Algal Res.* **2013**, *2*, 258–267. [CrossRef]

34. Wang, T.; Jónsdóttir, R.; Kristinsson, H.G.; Hreggvidsson, G.O.; Jónsson, J.Ó.; Thorkelsson, G.; Ólafsdóttir, G. Enzyme-enhanced extraction of antioxidant ingredients from red algae Palmaria palmata. *LWT-Food Sci. Technol.* **2010**, *43*, 1387–1393. [CrossRef]

35. Balasubramanian, S.; Allen, J.D.; Kanitkar, A.; Boldor, D. Oil extraction from Scenedesmus obliquus using a continuous microwave. *Bioresour. Technol.* **2011**, *102*, 3396–3403. [CrossRef] [PubMed]

36. Metherel, A.H.; Taha, A.Y.; Izadi, H.; Stark, K.D. The application of ultrasound energy to increase lipid extraction throughput of solid matrix samples (flaxseed). *Prostaglandins Leukot. Essent. Fatty Acids* **2009**, *81*, 417–423. [CrossRef] [PubMed]

37. Mustafa, A.; Turner, C. Pressurized liquid extraction as a green approach in food and herbal plants extraction: A review. *Anal. Chim. Acta* **2011**, *703*, 8–18. [CrossRef] [PubMed]

38. Andrich, G.; Nesti, U.; Venturi, F.; Zinnai, A.; Fiorentini, R. Supercritical fluid extraction of bioactive lipids from the microalga Nannochloropsis sp. *Eur. J. Lipid Sci. Technol.* **2005**, *107*, 381–386. [CrossRef]

39. Soha, L.; Zimmerman, J. Biodiesel production: The potential of algal lipids extracted with supercritical carbon dioxide. *Green Chem.* **2011**, *13*, 1422–1429. [CrossRef]

40. Patil, P.D.; Dandamudic, K.P.R.; Wanga, J.; Denga, Q.; Deng, S. Extraction of bio-oils from algae with supercritical carbon dioxide and co-solvents. *J. Supercrit. Fluids* **2018**, *135*, 60–68. [CrossRef]

41. Di Donna, L.; Benabdelkamel, H.; Mazzotti, F.; Napoli, A.; Nardi, M.; Sindona, G. High-throughput assay of oleopentanedialdheydes in extra virgin olive oil by the UHPLC– ESI-MS/MS and isotope dilution methods. *Anal. Chem.* **2011**, *83*, 1990–1995. [CrossRef]

42. Mazzotti, F.; Di Donna, L.; Taverna, D.; Nardi, M.; Aiello, D.; Napoli, A.; Sindona, G. Evaluation of dialdehydic anti-inflammatory active principles in extra-virgin olive oil by reactive paper spray mass spectrometry. *Int. J. Mass Spectrom.* **2013**, *352*, 87–91. [CrossRef]

43. Di Donna, L.; Mazzotti, F.; Santoro, I.; Sindona, G. Tandem mass spectrometry: A convenient approach in the dosage of steviol glycosides in Stevia sweetened commercial food beverages. *J. Mass. Spectrom.* **2017**, *52*, 290–295. [CrossRef] [PubMed]

44. Procopio, A.; Alcaro, S.; Nardi, M.; Oliverio, M.; Ortuso, F.; Sacchetta, P.; Pieragostino, P.; Sindona, G. Synthesis, biological evaluation, and molecular modeling of oleuropein and its semisynthetic derivatives as cyclooxygenase inhibitors. *JAFC* **2009**, *57*, 11161–11167. [CrossRef] [PubMed]

45. Procopio, A.; Celia, C.; Nardi, M.; Oliverio, M.; Paolino, D.; Sindona, G. Lipophilic hydroxytyrosol esters: Fatty acid conjugates for potential topical administration. *J. Nat. Prod.* **2011**, *74*, 2377–2381. [CrossRef] [PubMed]

46. Sindona, G.; Caruso, A.; Cozza, A.; Fiorentini, S.; Lorusso, B.; Marini, E.; Nardi, M.; Procopio, A.; Zicari, S. Anti-inflammatory effect of 3,4-DHPEA-EDA [2-(3,4-hydroxyphenyl) ethyl (3S, 4E)-4-formyl-3-(2-oxoethyl)hex-4-enoate] on primary human vascular endothelial cells. *Curr. Med. Chem.* **2012**, *19*, 4006–4013. [CrossRef] [PubMed]

47. Nardi, M.; Bonacci, S.; De Luca, G.; Maiuolo, J.; Oliverio, M.; Sindona, G.; Procopio, A. Biomimetic synthesis and antioxidant evaluation of 3, 4-DHPEA-EDA [2-(3,4-hydroxyphenyl) ethyl 4-formyl-3-(2-oxoethyl) hex-4-enoate]. *Food Chem.* **2014**, *162*, 89–93. [CrossRef] [PubMed]

48. Benincasa, C.; Perri, E.; Romano, E.; Santoro, I.; Sindona, G. Nutraceuticals from Olive Plain Water Extraction Identification and Assay by LC-ESI-MS/MS. *Anal. Bioanal. Chem.* **2015**, *6*, 1.

49. Nardi, M.; Bonacci, S.; Cariati, L.; Costanzo, P.; Oliverio, M.; Sindona, G.; Procopio, A. Synthesis and antioxidant evaluation of lipophilic oleuropein aglycone derivatives. *Food Funct.* **2017**, *8*, 4684–4692. [CrossRef]

50. Paonessa, R.; Nardi, M.; Di Gioia, M.L.; Olivito, F.; Oliverio, M.; Procopio, A. Eco-friendly synthesis of lipophilic EGCG derivatives and antitumor and antioxidant evaluation. *Nat. Prod. Commun.* **2018**, *13*, 1117–1122. [CrossRef]

51. Costanzo, P.; Bonacci, S.; Cariati, L.; Nardi, M.; Oliverio, M.; Procopio, A. Simple and efficient sustainable semi-synthesis of oleacein. [2-(3,4-hydroxyphenyl) ethyl (3S,4E)-4-formyl-3-(2-oxoethyl)hex-4-enoate] as potential additive for edible oils. *Food Chem.* **2018**, *245*, 410–414. [CrossRef]

52. Benincasa, C.; Santoro, I.; Nardi, M.; Cassano, A.; Sindona, G. Eco-Friendly Extraction and Characterisation of Nutraceuticals from Olive Leaves. *Molecules* **2019**, *24*, 3481. [CrossRef] [PubMed]

53. Clarke, D.; Ali, M.A.; Clifford, A.A.; Parratt, A.; Rose, P.; Schwinn, D.; Bannwarth, W.; Rayner, C.M. Reactions in unusual media. *Curr. Top Med. Chem.* **2004**, *4*, 729–771. [CrossRef] [PubMed]

54. Chiacchio, U.; Buemi, G.; Casuscelli, F.; Procopio, A.; Rescifina, A.; Romeo, R. Stereoselective synthesis of fused γ-lactams by intramolecular nitrone cycloaddition. *Tetrahedron* **1994**, *50*, 5503–5514. [CrossRef]

55. Bartoli, G.; Cupone, G.; Dalpozzo, R.; De Nino, A.; Maiuolo, L.; Marcantoni, E.; Procopio, A. Cerium-mediated deprotection of substituted allyl ethers. *Synlett* **2001**, *12*, 1897–1900. [CrossRef]

56. Procopio, A.; Cravotto, G.; Oliverio, M.; Costanzo, P.; Nardi, M.; Paonessa, R. An eco-sustainable erbium (iii)-catalyzed method for formation/cleavage of *O*-tert-butoxy carbonates. *Green Chem.* **2011**, *13*, 436–443. [CrossRef]

57. Procopio, A.; Costanzo, P.; Curini, M.; Nardi, M.; Oliverio, M.; Sindona, G. Erbium(III) chloride in ethyl lactate as a smart ecofriendly system for efficient and rapid stereoselective synthesis of trans-4,5-diaminocyclopent-2-enones. *ACS Sustain. Chem. Eng.* **2013**, *1*, 541–544. [CrossRef]

58. Nardi, M.; Cano, N.H.; Costanzo, P.; Oliverio, M.; Sindona, G.; Procopio, A. Aqueous MW eco-friendly protocol for amino group protection. *RSC Adv.* **2015**, *5*, 18751–18760. [CrossRef]

59. Nardi, M.; Sindona, G.; Costanzo, P.; Oliverio, M.; Procopio, A. Eco-friendly stereoselective reduction of α,β-unsaturated carbonyl compounds by Er(OTf)$_3$/NaBH$_4$ in 2-MeTHF. *Tetrahedron* **2015**, *71*, 1132–1135. [CrossRef]

60. Oliverio, M.; Nardi, M.; Cariati, L.; Vitale, E.; Bonacci, S.; Procopio, A. "on Water" MW-Assisted Synthesis of Hydroxytyrosol Fatty Esters. *ACS Sustain. Chem. Eng.* **2016**, *4*, 661–665. [CrossRef]

61. Nardi, M.; Costanzo, P.; De Nino, A.; Di Gioia, M.L.; Olivito, F.; Sindona, G.; Procopio, A. Water excellent solvent for the synthesis of bifunctionalized cyclopentenones from furfural. *Green Chem.* **2017**, *19*, 5403–5411. [CrossRef]

62. Nardi, M.; Di Gioia, M.L.; Costanzo, P.; De Nino, A.; Maiuolo, L.; Oliverio, M.; Olivito, F.; Procopio, A. Selective acetylation of small biomolecules and their derivatives catalyzed by Er(OTf)$_3$. *Catalysts* **2017**, *7*, 269. [CrossRef]

63. Di Gioia, M.L.; Nardi, M.; Costanzo, P.; De Nino, A.; Maiuolo, L.; Oliverio, M.; Procopio, A. Biorenewable deep eutectic solvent for selective and scalable conversion of furfural into cyclopentenone derivatives. *Molecules* **2018**, *23*, 1891. [CrossRef] [PubMed]

64. Di Gioia, M.L.; Cassano, R.; Costanzo, P.; Herrera Cano, N.; Maiuolo, L.; Nardi, M.; Nicoletta, F.P.; Oliverio, M.; Procopio, A. Green Synthesis of Privileged Benzimidazole Scaffolds Using Active Deep Eutectic Solvent. *Molecules* **2019**, *24*, 2885. [CrossRef] [PubMed]

65. Watanabe, K.; Yamagiwa, N.; Torisawa, Y. Cyclopentyl methyl ether as a new and alternative process solvent. *Org. Process Res. Dev.* **2007**, *11*, 251–258. [CrossRef]

66. Antonucci, V.; Coleman, J.; Ferry, J.B.; Johnson, N.; Mathe, M.; Scott, J.P.; Xu, J. Toxicological assessment of 2-methyltetrahydrofuran and cyclopentyl methyl ether in support of their use in pharmaceutical chemical process development. *Org. Process Res. Dev.* **2011**, *15*, 939–941. [CrossRef]

67. Campos Molina, M.J.; Mariscal, R.; Ojeda, M.; Lopez Granados, M. Cyclopentyl methyl ether: A green co-solvent for the selective dehydration of lignocellulosic pentoses to furfural. *Bioresour. Technol.* **2012**, *126*, 321–327. [CrossRef]

68. Yara-Varò, E.; Fabiano-Tixier, A.S.; Balcells, M.; Canela-Garayoa, R.; Bily, A.; Chemat, F. Is it possible to substitute hexane with green solvents for extraction of carotenoids? A theoretical versus experimental solubility study. *RSC Adv.* **2016**, *6*, 58055–58063.

69. Wales, M.D.; Huang, C.; Joos, L.B.; Probst, K.V.; Vadlani, P.V.; Anthony, J.L.; Rezac, M.E. Liquid–liquid equilibria for ternary systems of water/methoxycyclopentane/alcohol (methanol, ethanol, 1-propanol, or 2-propanol). *J Chem. Eng. Data.* **2016**, *61*, 1479–1484. [CrossRef]

70. De Jesus, S.S.; Ferreira, G.F.; Moreira, L.S.; Wolf Maciel, M.R.; Maciel Filho, R. Comparison of several methods for effective lipid extraction from wet microalgae using green solvents. *Renew. Energy* **2019**, *143*, 130–141. [CrossRef]

71. Danielewicz, M.A.; Anderson, L.A.; Franz, A.K. Triacylglycerol profiling of marine microalgae by mass spectrometry. *J. Lipid Res.* **2011**, *52*, 2101–2108. [CrossRef]

72. Jones, J.; Manning, S.; Montoya, M.; Keller, K.; Poenie, M. Extraction of algal lipids and their analysis by HPLC and mass spectrometry. *J. Am. Oil Chem. Soc.* **2012**, *89*, 1371–1381. [CrossRef]

73. Laakso, P. Mass spectrometry of triacylglycerols. *Eur. J. Lipid Sci. Technol.* **2002**, *104*, 43–49. [CrossRef]

74. Atolani, O.; Olabiyi, E.T.; Issa, A.A.; Azeez, H.T.; Onoja, E.G.; Ibrahim, S.O.; Zubair, M.F.; Oguntoye, O.S.; Olatunji, G.A. Green synthesis and characterisation of natural antiseptic soaps from the oils of underutilised tropical seed. *Sustain. Chem. Pharm.* **2016**, *4*, 32–39. [CrossRef]

75. Das, U.N.; Fams, M.D. Long-chain polyunsaturated fatty acids in the growth and development of the brain and memory. *Nutrition* **2003**, *19*, 62–65. [CrossRef]

76. Nettleton, J.A. Are n-3 fatty acids essential nutrients for fetal and infant development? *J. Am. Diet. Assoc.* **1992**, *93*, 58–64. [CrossRef]

77. Mendes, A.; Lopes da Silva, T.; Reis, A. DHA Concentration and Purification from the Marine Heterotrophic Microalga *Crypthecodinium cohnii* CCMP 316 by Winterization and Urea Complexation. *Food Technol. Biotechnol.* **2007**, *45*, 38–44.

78. Shahidi, F.; Wanasundara, U. Omega-3 fatty acid concentrates: Nutritional aspects and production technologies. *Food Sci. Technol.* **1998**, *9*, 230–240. [CrossRef]

79. Shahidi, F.; Senanayake, S.P.J.N. Concentration of docosahexaenoic acid (DHA) from algal oil via urea complexation. *J. Food Lipids* **2000**, *7*, 51–61.

80. Dunn, P.J. The importance of Green Chemistry in Process Research and Development. *Chem. Soc. Rev.* **2012**, *41*, 1452–1461. [CrossRef]

81. Jeevan, K.S.P.; Garlapati, V.K.; Dash, A.; Scholz, P.; Banerjee, R. Sustainable green solvents and techniques for lipid extraction from microalgae: A review. *Algal Res.* **2017**, *21*, 138–147. [CrossRef]

82. Ecker, J.; Scherer, M.; Schmitz, G.; Liebisch, G. A rapid GC-MS method for quantification of positional and geometric isomers of fatty acid methyl esters. *J. Chromatogr. B: Anal. Technol. Biomed. Life Sci.* **2012**, *897*, 98–104. [CrossRef] [PubMed]

83. Akoto, L.; Vreuls, R.J.J.; Irth, H.; Pel, R.; Stellaard, F. Fatty acid profiling of raw human plasma and whole blood using direct thermal desorption combined with gas chromatography–mass spectrometry. *J. Chromatogr. A* **2008**, *1186*, 365–371. [CrossRef] [PubMed]

84. Nagy, K.; Jakab, A.; Fekete, J.; Vekey, K. An HPLC-MS approach for analysis of very long chain fatty acids and other apolar compounds on octadecyl-silica phase using partly miscible solvent. *Anal. Chem.* **2004**, *76*, 1935–1941. [CrossRef] [PubMed]
85. Murphy, R.C.; Fiedler, J.; Hevko, J. Analysis of non-volatile lipids by mass spectrometry. *Chem. Rev.* **2001**, *101*, 479–526. [CrossRef]
86. Kerwin, J.L.; Wiens, A.M.; Ericsson, L.H. Identification of fatty acids by electrospray mass spectrometry and tandem mass spectrometry. *J. Mass Spectrom.* **1996**, *31*, 184–192. [CrossRef]
87. Malik, A.K.; Blasco, C.; Pico, Y. Liquid chromatography-mass spectrometry in food safety. *J. Chromatogr. A* **2010**, *1217*, 4018–4040. [CrossRef]

Sample Availability: Samples of the compounds are not available from the authors.

Article

Facile Preparation of CuS Nanoparticles from the Interfaces of Hydrophobic Ionic Liquids and Water

Yunchang Fan [1] , Yingcun Li [1], Xiaojiang Han [2], Xiaojie Wu [1], Lina Zhang [1,*] and Qiang Wang [1,*]

[1] College of Chemistry and Chemical Engineering, Henan Polytechnic University, Jiaozuo 454003, China; fanyunchang@hpu.edu.cn (Y.F.); m13080151080@sohu.com (Y.L.); latiaowu@sina.com (X.W.)

[2] Zhenhai District Center for Disease Control and Prevention, Ningbo 315200, China; zjsxhxj@163.com

* Correspondence: zhln@hpu.edu.cn (L.Z.); wangqiang@hpu.edu.cn (Q.W.); Tel.: +86-391-398-6813 (L.Z.)

Academic Editors: Monica Nardi, Antonio Procopio and Maria Luisa Di Gioia
Received: 17 September 2019; Accepted: 19 October 2019; Published: 21 October 2019

Abstract: In this work, a two-phase system composed of hydrophobic ionic liquid (IL) and water phases was introduced to prepare copper sulfide (CuS) nanoparticles. It was found that CuS particles generated from the interfaces of carboxyl-functionalized IL and sodium sulfide (Na_2S) aqueous solution were prone to aggregate into nanoplates and those produced from the interfaces of carboxyl-functionalized IL and thioacetamide (TAA) aqueous solution tended to aggregate into nanospheres. Both the CuS nanoplates and nanospheres exhibited a good absorption ability for ultraviolet and visible light. Furthermore, the CuS nanoplates and nanospheres showed highly efficient photocatalytic activity in degrading rhodamine B (RhB). Compared with the reported CuS nanostructures, the CuS nanoparticles prepared in this work could degrade RhB under natural sunlight irradiation. Finally, the production of CuS from the interfaces of hydrophobic IL and water phases had the advantages of mild reaction conditions and ease of operation.

Keywords: ionic liquids (ILs); copper sulfide (CuS); nanoparticles; photocatalytic activity

1. Introduction

Copper sulfide (CuS), a p-type semiconductor, has attracted tremendous interest due to its excellent optical and electronic properties. The optical band gap energy of CuS depends on its crystalline phase and is in the range of 1.48–2.89 eV, which matches the energy of ultraviolet and visible light (4.1–1.6 eV, 300–800 nm) [1,2], meaning that CuS has a strong absorption ability for ultraviolet and visible light and can be widely used in many fields, such as for the photocatalytic degradation of organic pollutants, solar cells, optical filters, and superconductors [3–5]. In the photocatalytic degradation of organic pollutants, CuS is a Fenton-like catalyst, a type of catalyst that effectively decomposes a wide range of organic pollutants in the presence of hydrogen peroxide (H_2O_2) with light [4,5]. Like other Fenton-like catalysts, CuS catalyzes the decomposition of H_2O_2 to generate a large number of hydroxyl radicals ($^\bullet OH$) and superoxide ions ($^\bullet O_2^-$) under light irradiation. Subsequently, $^\bullet OH$ and $^\bullet O_2^-$ degrade the organic pollutants. The photocatalytic degradation of organic pollutants using CuS as a catalyst can be described by the following equations [4–6]:

$$CuS + light\ irradiation \rightarrow h^+\ (CuS) + e^-\ (CuS) \tag{1}$$

$$h^+\ (CuS) + H_2O_2 \rightarrow {}^\bullet OOH + H^+ \tag{2}$$

$$e^-\ (CuS) + H_2O_2 \rightarrow {}^\bullet OH + OH^- \tag{3}$$

$${}^\bullet OOH \rightarrow {}^\bullet O_2^- + H^+ \tag{4}$$

$$Organic\ pollutants + {}^\bullet OH + {}^\bullet O_2^- \rightarrow degradation\ products \tag{5}$$

where e^- (CuS) and h^+ (CuS) are the electrons jumping from the valence band to the conduction band of CuS and the positively charged holes in the valence band of CuS, respectively. The symbol $^\bullet$OOH refers to the hydroperoxy radical.

A large amount of research has suggested that the size reduction of CuS into the nanoscale dimension leads to significant changes in its physicochemical properties due to the quantum size effect [7–9]. Therefore, in the last few years, various techniques, such as the hydrothermal method, microwave irradiation, and electrodeposition have been developed to prepare different shapes for these CuS nanoparticles like nanospheres, nanoplates, nanotubes, nanorods, and flower-like structures [5,7–12]. Among these techniques, the hydrothermal method is the most commonly used one in the synthesis of CuS nanoparticles. For example, Cheng et al. introduced a hydrothermal method (reaction temperature, 120–180 °C; reaction time, 6–24 h) to synthesize ball-flower shaped CuS nanoparticles by mixing the solutions of copper chloride and thiourea using poly(vinylpyrrolidone) (PVP) as the surfactant. The morphologies of ball-flower shaped CuS are highly dependent on the concentration ratios of copper chloride to thiourea, the reaction temperature, and reaction time. Finally, this ball-flower shaped CuS shows high photocatalytic activity for the degradation of rhodamine B (RhB) under exposure to ultraviolet light irradiation [9]. Mezgebe and coworkers suggested a hydrothermal method to prepare nano-sized CuS by heating the solution containing $Cu(NO_3)_2$ and thiourea at 150 °C for 6 h. The different morphological structures of CuS were prepared by changing its hydrothermal solvents. The prepared nano-sized CuS showed high catalytic activity for the degradation of its model pollutant, methylene blue [11]. Generally, hydrothermal solvents have a significant effect on the morphology and activity of nano-sized CuS [5,11] and thus the development of versatile solvents is a research hotspot. In this context, the use of new type of solvent, ionic liquids (ILs), to hydrothermally produce nano-sized CuS has aroused great interest among researchers [13–16] because ILs have some unique properties, such as high viscosity and good solubility for a wide range of inorganic and organic compounds and their physicochemical properties can be easily adjusted by changing the chemical structures of their anions and cations [17,18]. Yao et al. reported the solvothermal synthesis of CuS nanowalls in the presence of a common IL, 1-dedyl-3-methylimidazolium bromide ([C_{10}mim]Br) (heating at 150 °C for 10 h). This IL may be involved in the formation of the thiourea-Cu(II) complex via surface absorption or as one of elements in the complex, which can effectively modulate the morphology of CuS. Furthermore, the chain length of the IL cation, the IL anion's nature, and the IL imidazolium ring also played an important role in delicately constructing the CuS nanowalls [13].

From the above discussion, there is no doubt that hydrothermal method with ILs as solvents is an ideal technique for the preparation of nano-sized CuS. However, this method usually operates at higher temperature, takes a longer time and does not maximize the effectiveness of ILs because the ILs used in the reported work only act as additives, which cannot reflect the advantages of ILs (such as their excellent solubility and higher viscosity). To make full use of the advantages of ILs, this work suggests a biphasic system composed of a hydrophobic IL phase that contains copper ions and a water phase that contains sulfide ions. The CuS nanoparticles would be generated in the interfaces between the two phases. The solubility of copper ions in the IL phase suggests the slow release of copper ions in the phase interfaces, and the high viscosity of the ILs suggests the slow diffusion of nanoparticles in the IL phase, which may facilitate the orderly growth of nanoparticles. Based on this conception, this work designed a biphasic system consisting of hydrophobic IL and water phases to produce CuS nanoparticles and investigated their photocatalytic activities.

2. Results and Discussion

In this work, CuS nanoparticles were prepared using a simpler route which was based on a two-phase system composed of hydrophobic ILs (1-octyl-3-methylimidazolium bis(trifluoromethylsulfonyl)imide ([C_8mim]NTf$_2$) and 1-butyl-3-carboxymethylimidazolium bis(trifluoromethylsulfonyl)imide ([C_4C_2OOHim]NTf$_2$)) and water with copper(II) acetate (Cu(Ac)$_2$) as a copper source and sodium sulfide (Na$_2$S) and thioacetamide (TAA) as sulfur sources. Compared with the hydrothermal method (usually operating at 120–180 °C [9,11,13]), a traditional technique to produce CuS nanoparticles, the developed hydrophobic IL/water systems were actuated at mild conditions (the IL/Na$_2$S systems were carried out at room temperature, and the IL/TAA systems were conducted at 80 °C). The results discussed below indicate that the prepared CuS nanostructures have a pronounced absorption ability for ultraviolet and visible light and exhibit good photocatalytic efficiency toward the degradation of the virulent organic pollutant, RhB.

2.1. Morphology and Structural Characterization of CuS

The crystal structures and phase purity of the CuS prepared by different systems were investigated by X-ray diffraction (XRD). The results are shown in Figures 1 and 2. As can be seen from the two figures, for the CuS nanoparticles prepared from the interfaces of the [C$_8$mim]NTf$_2$/water systems and the [C$_4$C$_2$OOHim]NTf$_2$/TAA system (80 °C, 2 h), there are some impurity peaks in the XRD spectra, which are the characteristic signals of CuSO$_4$•5H$_2$O. Meanwhile, a higher purity of CuS can be obtained when the [C$_4$C$_2$OOHim]NTf$_2$/Na$_2$S system is used as the reaction medium, which may ascribed to its acidic nature and strong complexation ability of the carboxyl group in the [C$_4$C$_2$OOHim]NTf$_2$ with Cu^{2+}. As shown in Figure 2, for the [C$_4$C$_2$OOHim]NTf$_2$/TAA system, the heating time is also a key factor that affects the product's purity. A shorter heating time (2 h) leads to small amounts of impurities in the product, and there are no impurity peaks found in the XRD spectra when the heating times were set at 4 h and 6 h.

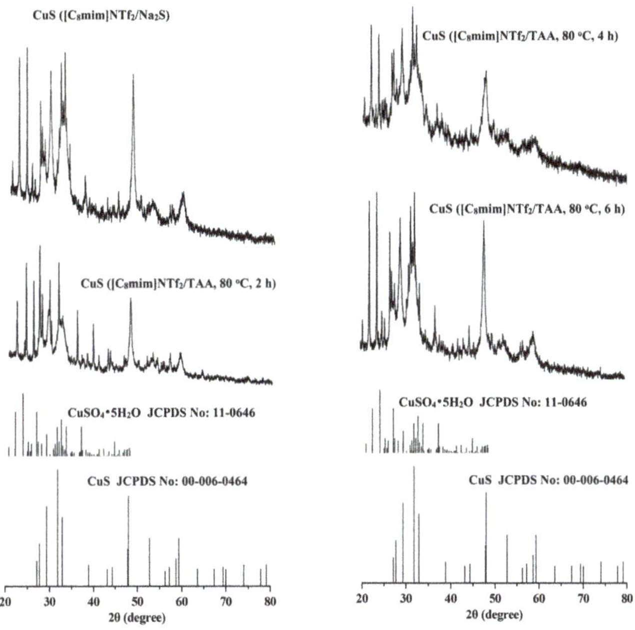

Figure 1. X-ray diffraction (XRD) patterns of the copper sulfide (CuS) prepared in the [C$_8$mim]NTf$_2$/water systems.

Since the [C$_4$C$_2$OOHim]NTf$_2$/Na$_2$S system and the [C$_4$C$_2$OOHim]NTf$_2$/TAA system (80 °C, 4 h or 6 h) can provide high purity products, they were selected for the following studies. The transmission electron microscopic (TEM) and field-emission scanning electron microscopic (FE-SEM) images of CuS obtained from the [C$_4$C$_2$OOHim]NTf$_2$/Na$_2$S and [C$_4$C$_2$OOHim]NTf$_2$/TAA systems are shown in Figures 3 and 4. As can be seen from Figure 3, the TEM images of the as-prepared CuS clearly show the formation of plate-like nanostructures and the particles tend to agglomerate to some extent. The SEM images (Figure 4) of CuS indicate that the CuS nanoparticles obtained from the [C$_4$C$_2$OOHim]NTf$_2$/Na$_2$S system self-assembled to form large plate structures and the CuS nanoparticles generated from [C$_4$C$_2$OOHim]NTf$_2$/TAA systems are prone to form rough and spheroidic structures.

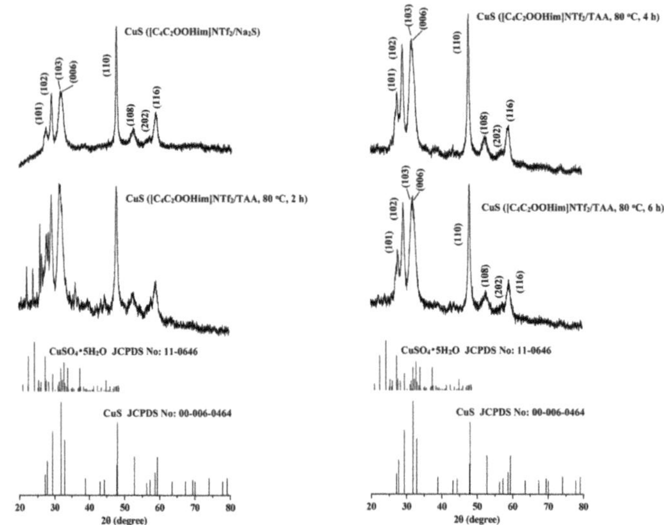

Figure 2. XRD patterns of the CuS prepared in the [C$_4$C$_2$OOHim]NTf$_2$/water systems.

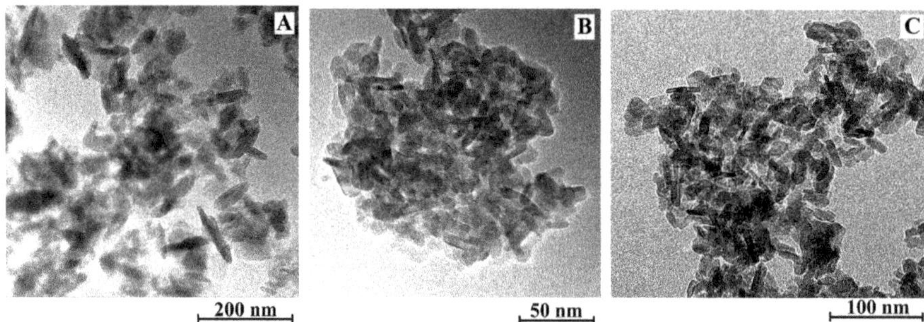

Figure 3. Transmission electron microscopic (TEM) images of the CuS prepared in different systems. (**A**) [C$_4$C$_2$OOHim]NTf$_2$/Na$_2$S, (**B**) [C$_4$C$_2$OOHim]NTf$_2$/TAA (80 °C, 4 h) and (**C**) [C$_4$C$_2$OOHim]NTf$_2$/TAA (80 °C, 6 h).

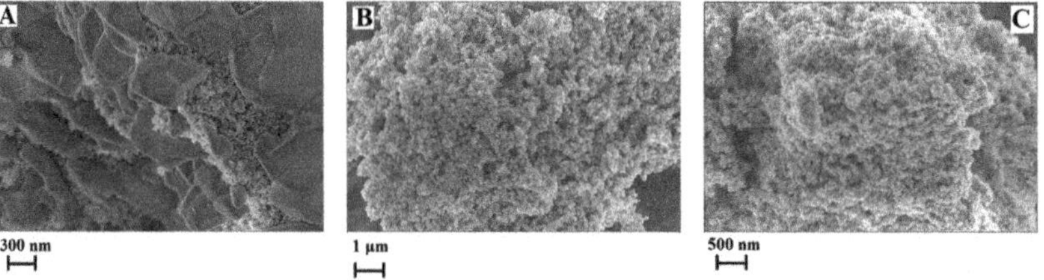

Figure 4. Field-emission scanning electron microscopic (FE-SEM) images of the CuS prepared in different systems. (**A**) [C$_4$C$_2$OOHim]NTf$_2$/Na$_2$S, (**B**) [C$_4$C$_2$OOHim]NTf$_2$/TAA (80 °C, 4 h) and (**C**) [C$_4$C$_2$OOHim]NTf$_2$/TAA (80 °C, 6 h).

Additionally, the specific surface areas of the as-prepared CuS nanoparticles were analyzed by the Brunauer-Emmett-Teller (BET) method. It was found that the specific surface areas of the prepared CuS nanoparticles are 45.4 m^2 g^{-1} ([C$_4$C$_2$OOHim]NTf$_2$/Na$_2$S system), 28.7 m^2 g^{-1} ([C$_4$C$_2$OOHim]NTf$_2$/TAA system, 80 °C, 4 h), and 18.3 m^2 g^{-1} ([C$_4$C$_2$OOHim]NTf$_2$/TAA system, 80 °C, 6 h), respectively, which are larger than those of the hydrothermally prepared CuS nanoplates [11].

2.2. Ultraviolet–Visible (UV/Vis) Spectra of CuS

The UV/Vis absorption spectra of the resulting CuS products are shown in Figure 5. It is clear that all the spectra show strong absorption feature in the UV/Vis region, which suggests the potential application of CuS in the fields of solar cells and photocatalysts. Moreover, the absorption peak of the CuS prepared from the [C$_4$C$_2$OOHim]NTf$_2$/Na$_2$S system (maximum absorption wavelength is around 582 nm) show a blue shift compared with the absorption peaks of the CuS produced from the [C$_4$C$_2$OOHim]NTf$_2$/TAA system (maximum absorption wavelengths are about 679 nm), implying that the CuS nanoplates synthesized from the [C$_4$C$_2$OOHim]NTf$_2$/Na$_2$S system are more uniform in size and morphology (Figure 4) [19].

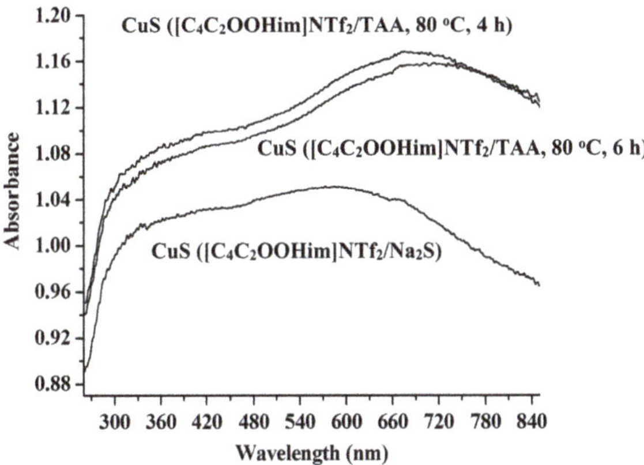

Figure 5. UV/Vis spectra of the CuS nanoparticles prepared in different systems.

2.3. Catalytic Activities

It has been reported that CuS nanoparticles are good photocatalysts for the degradation of organic dyes, such as RhB [6,9,20–23]. Herein, the photocatalytic performance of the as-synthesized CuS nanostructures for the degradation of RhB in the presence of H$_2$O$_2$ under natural light irradiation at ambient temperature was investigated. The results shown in Figure 6 illustrate that all the as-synthesized CuS nanoparticles exhibit high photocatalytic activity and over 90% of the RhB is decomposed after 60 min. To further characterize the catalytic performance of the prepared CuS, the relationship between ln (C$_0$/C) and degradation time (min) was studied and the results shown in Figure 7 suggest that the RhB degradation conforms to the first-order kinetic model, which is in good agreement with the observations reported in the literature [6,21,24]. The RhB degradation kinetics can be expressed by the following equation:

$$\ln(C_0/C_i) = kt \tag{6}$$

where k refers to the reaction rate constant and C_0 and C_i are the initial and instant RhB concentrations, respectively.

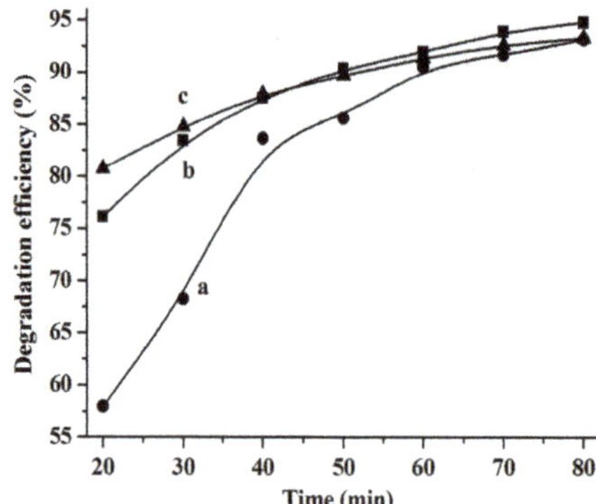

Figure 6. The rhodamine B (RhB) degradation efficiency catalyzed by the CuS produced from different systems. (a) [C$_4$C$_2$OOHim]NTf$_2$/TAA (80 °C, 4 h), (b) [C$_4$C$_2$OOHim]NTf$_2$/Na$_2$S, and (c) [C$_4$C$_2$OOHim]NTf$_2$/TAA (80 °C, 6 h); 5 mg of CuS, 25 mL of RhB (10 mg L^{-1}), and 1.0 mL of H$_2$O$_2$ (30 wt.%).

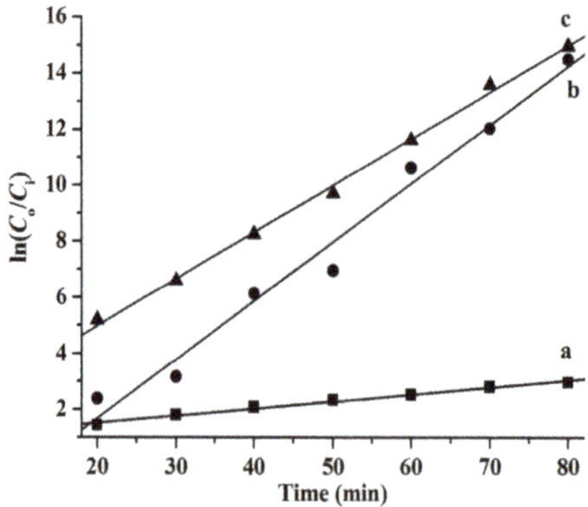

Figure 7. Degradation kinetics of RhB catalyzed by the CuS synthesized from different systems. (a) [C$_4$C$_2$OOHim]NTf$_2$/Na$_2$S, (b) [C$_4$C$_2$OOHim]NTf$_2$/TAA (80 °C, 4 h), and (c) [C$_4$C$_2$OOHim]NTf$_2$/TAA (80 °C, 6 h); 5 mg of CuS, 25 mL of RhB (10 mg L^{-1}), and 1.0 mL of H$_2$O$_2$ (30 wt.%).

It was found that the reaction rate constants (k values) of the RhB degradation catalyzed by CuS are 0.025 min^{-1} ([C$_4$C$_2$OOHim]NTf$_2$/Na$_2$S system), 0.209 min^{-1} ([C$_4$C$_2$OOHim]NTf$_2$/TAA system, 80 °C, 4 h), and 0.167 min^{-1} ([C$_4$C$_2$OOHim]NTf$_2$/TAA system, 80 °C, 6 h), respectively.

In addition, the dosage of H$_2$O$_2$ is one of key factors affecting the degradation efficiency of organic dyes [6,23]. Therefore, the influence of H$_2$O$_2$ dosage on the degradation efficiency was investigated, and the results shown in Figure 8A indicate that the degradation efficiency of the CuS nanoplates obtained from the [C$_4$C$_2$OOHim]NTf$_2$/Na$_2$S system increases as the H$_2$O$_2$ dosage increases from 0.59% (wt.%) to 1.7% and remains constant with an increasing H$_2$O$_2$ dosage up to 2.2%. That is to say, 1.7% of H$_2$O$_2$ is a better choice for the degradation of RhB by CuS nanoplates. As illustrated in Figure 8B, the degradation efficiency of the CuS nanospheres obtained from the [C$_4$C$_2$OOHim]NTf$_2$/TAA system (80 °C, 4 h) also increases by increasing the H$_2$O$_2$ dosage from 0.59% to 2.2% and when 2.2% of H$_2$O$_2$ is adopted, the degradation efficiency of RhB increases up to 92.4% after 40 min of reaction. Thus, 2.2% is regarded as the optimal H$_2$O$_2$ dosage for the CuS nanopheres ([C$_4$C$_2$OOHim]NTf$_2$/TAA, 80 °C, 4 h). For the CuS nanospheres produced from the [C$_4$C$_2$OOHim]NTf$_2$/TAA system (80 °C, 6 h), their catalytic capacity also increases by increasing the H$_2$O$_2$ dosage from 0.59% to 2.2% (Figure 8C), and the degradation efficiency goes above 90% when 1.7% of H$_2$O$_2$ is used after 30 min. Thus, 1.7% is regarded as the optimal H_2O_2 dosage for the CuS nanopheres obtained from the [C$_4$C$_2$OOHim]NTf$_2$/TAA system (80 °C, 6 h).

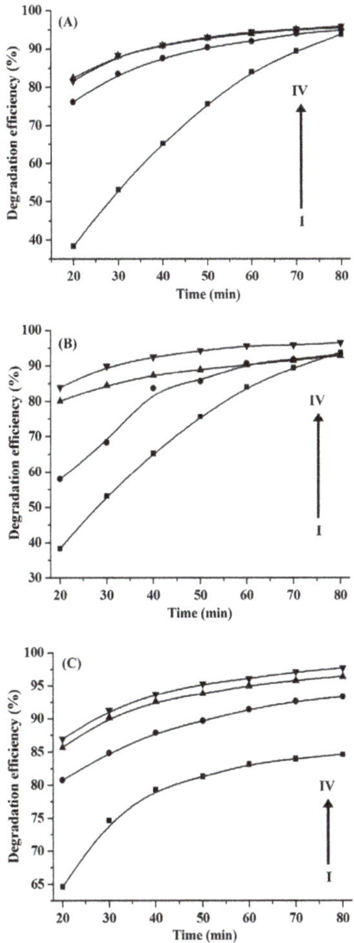

Figure 8. The influence of H_2O_2 dosage on the degradation efficiency of the RhB catalyzed by CuS prepared from different systems. (**A**) [C$_4$C$_2$OOHim]NTf$_2$/Na$_2$S system, (**B**) [C$_4$C$_2$OOHim]NTf$_2$/TAA system (80 °C, 4 h) and (**C**) [C$_4$C$_2$OOHim]NTf$_2$/TAA system (80 °C, 6 h); 5 mg of CuS, 25 mL of RhB (10 mg L^{-1}); from I to IV, 0.59% of H_2O_2, 1.2% of H_2O_2, 1.7% of H_2O_2, and 2.2% of H_2O_2, respectively.

Recently, some excellent work has reported the use of CuS nanostructures with different morphologies to decompose RhB [6,9,20,21,23]. Therefore, a comparison of the photocatalytic performance between the CuS nanoparticles synthesized in the present work and those reported in the literature was conducted, and the results are listed in Table 1. As can be seen, the degradation of RhB by CuS hollow nanospheres [20], nanoneedles [21], nanospheres [23] and ball-flowers [9] was conducted under an extra light source, such as Xe and mercury lamps. However, the RhB degradation reaction catalyzed by the CuS synthesized in this work can be carried out under natural sunlight. This is an advantage of the as-prepared CuS nanoparticles. Furthermore, more than 90% of RhB was degraded after 30–40 min of reaction, suggesting that the as-prepared CuS has a stronger photocatalytic capacity.

Table 1. Comparison of the photocatalytic performance of CuS nanoparticles prepared by this work with reported CuS nanostructures.

Catalyst	Mass Ratio of RhB to Catalyst (m_{RhB}/m_{CuS})	Degradation Time (min)	Irradiation Source	Degradation Efficiency (%, v/v)	Reference
Flower-like CuS hollow nanospheres	10^{-2}	40	Xe lamp (150 W)	>90%	20
CuS nanoneedles	1.25×10^{-2}	105	Mercury lamp (500 W)	91%	21
CuS hierarchical microflowers	6.0×10^{-2}	55	Natural light	95%	6
CuS nanospheres	2.0×10^{-2}	20	Xe lamp (300 W)	About 100%	23
CuS ball-flowers	4.8×10^{-2}	60	Mercury lamp (300 W)	About 100%	9
CuS nanoplates	5.0×10^{-2}	40	Natural light	90.8%	This work
CuS nanospheres	5.0×10^{-2}	30	Natural light	90.2%	This work

Finally, it should be noted that CuS nanoparticles have good biocompatibility [25,26] and the ILs used in this work have very low volatility, whereby any air pollution caused by them can be significantly reduced [27]. Furthermore, after reaction, the ILs can be recycled through washing with HCl (0.2 mol L^{-1}) and drying (60 °C for 4 h) processes. Therefore, the suggested IL/water systems are environmentally friendly methods for the preparation of CuS nanoparticles.

3. Materials and Methods

3.1. Materials

Copper(II) acetate monohydrate (Cu(Ac)$_2$·H$_2$O, 99%), thioacetamide (TAA, ≥99%), 1-methylimidazole (MI, 99%), rhodamine B (RhB, 99%), and hydrogen peroxide solution (H$_2$O$_2$, 30 wt.% in water) were obtained from Aladdin Chemical Co., Ltd. (Shanghai, China). The ILs, 1-octyl-3-methylimidazolium bis(trifluoromethylsulfonyl)imide ([C$_8$mim]NTf$_2$, 99%) and 1-butyl-3-carboxymethylimidazolium bis(trifluoromethylsulfonyl)imide ([C$_4$C$_2$OOHim]NTf$_2$, 95%) were purchased from Chengjie Chemical Co., Ltd. (Shanghai, China). Unless otherwise stated, all other reagents used were of analytical grade. Ultrapure water produced by an Aquapro purification system (18.2 MΩ cm, Aquapro International Co., Ltd., Dover, DE, USA) was used throughout the experiments.

3.2. Preparation of Nano CuS

The preparation of the nano CuS in the [C$_8$mim]NTf$_2$ + MI/water system was as follows: 11 mL of [C$_8$mim]NTf$_2$ (containing 9.1% (volume percentage) of MI) and 20 mL of Cu(Ac)$_2$ solution (0.1 mol L^{-1}) were mixed under stirring at room temperature for 30 min. Here, MI acted as a complexant, i.e., it could form complexes with copper ions, resulting in the transfer of copper ions to the IL phase. The water phase was discarded and the IL phase was washed with water to remove the undissolved Cu(Ac)$_2$, and then Na$_2$S (22 mL, 0.1 mol L^{-1}) was added dropwise to the IL phase under stirring for 15 min. The color of the IL phase changed from blue to black, indicating the formation of CuS. The IL phase was dissolved with ethanol to precipitate CuS. The products were washed several times with ethanol, followed by washing with water and acetone, the resultant CuS was dried at 40 °C for 2 h.

When TAA was used as a sulfur source, the preparation of CuS followed a similar pattern, except that TAA (0.1 mol L^{-1}, prepared with 0.12 mol L^{-1} of HCl solution) was used instead of Na$_2$S, and the reaction was conducted at 80 °C [28].

Preparation of the nano CuS in [C$_4$C$_2$OOHim]NTf$_2$ followed a similar procedure (as described above), except [C$_4$C$_2$OOHim]NTf$_2$ (without the addition of MA or HCl) was used instead of [C$_8$mim]NTf$_2$.

All CuS samples were characterized by an X-ray diffractometer (XRD, model X'Pert PRO MPD, PANalytical B.V., Almelo, Netherlands), a field-emission scanning electron microscope (FE-SEM, Quanta 250 FEG, Thermo Fisher Scientific, Hillsboro, OR, USA), a transmission electron microscope (TEM, model Tecnai G2 20, FEI, Hillsboro, OR, USA), a surface area and porosity analyzer (model ASAP 2460, Micromeritics Instrument Corp., Norcross, GA, USA) and an ultraviolet-visible/near infrared spectrophotometer (Model UH4150, Hitachi High-Technologies Corp., Tokyo, Japan).

3.3. Measurement of Photocatalytic Activity

The photocatalytic activity of the prepared CuS nanoparticles was evaluated by degrading the model compound, RhB, in aqueous media under natural sunlight irradiation. The experiments were conducted according to the procedures reported in the literature [6,9,20,21]. Typically, 5.0 mg of CuS nanoparticles was added in the RhB solution (25 mL, 10 mg L^{-1}) and the resultant reaction mixture was magnetically stirred in the dark for 30 min to obtain adsorption–desorption equilibrium. After that, a certain amount of H_2O_2 (30 wt.%) was added into the reaction mixture, and the photocatalytic degradation of RhB was initiated by natural light irradiation. The RhB concentration in the reaction mixture was measured by an ultraviolet-visible (UV/Vis) spectrophotometer (model TU-1810, Purkinje General Instrument Co., Beijing, China) at 550 nm. The degradation efficiency was calculated by using the following equation:

$$\text{Degradation efficiency} = (1 - C_i/C_o) \times 100\% \tag{7}$$

where C_o and C_i are the initial and instant RhB concentrations, respectively.

4. Conclusions

The present work has suggested a facile method to prepare CuS nanoparticles from the interfaces of hydrophobic IL and water phases. The morphologies of CuS could be selectively produced by changing the types of their sulfur sources. The use of Na_2S led to the production of CuS nanoplates and CuS nanospheres could be obtained when TAA is used as a sulfur source. All the prepared CuS nanostructures exhibited strong UV/Vis absorption and excellent photocatalytic activities toward the degradation of RhB under exposure to natural sunlight irradiation. Compared with the reported techniques, the suggested two-phase method exhibited some advantages for the preparation of CuS, such as mild reaction conditions and simplified reaction procedures. This facile and eco-friendly strategy is expected to produce other metal sulfide nano-structures with great promise for various applications.

Author Contributions: Y.F., conceptualization and writing—review and editing; Y.L., investigation; X.H., data curation; X.W., writing—original draft preparation; L.Z., conceptualization; Q.W., methodology.

Funding: This research was financially supported by the National Natural Science Foundation of China (21805071), Foundation of Henan province (No. 182102310728), the Program for Innovative Research Team of Henan Polytechnic University (No. T2018-3) and the Fundamental Research Funds for the Universities of Henan Province (NSFRF180415).

Conflicts of Interest: The authors declare no conflict of interest.

References

1. Zhao, L.; Zhou, L.; Sun, C.; Gu, Y.; Wen, W.; Fang, X. Rose-like CuS microflowers and their enhanced visible-light photocatalytic performance. *CrystEngComm* **2018**, *20*, 6529–6537. [CrossRef]
2. Riyaz, S.; Parveen, A.; Azam, A. Microstructural and optical properties of CuS nanoparticles prepared by sol-gel route. *Perspect. Sci.* **2016**, *8*, 632–635. [CrossRef]
3. Saranya, M.; Santhosh, C.; Augustine, S.P.; Grace, A.N. Synthesis and characterisation of CuS nanomaterials using hydrothermal route. *J. Exp. Nanosci.* **2014**, *9*, 329–336. [CrossRef]
4. Nethravathi, C.; Nath, R.R.; Rajamathi, J.T.; Rajamathi, M. Microwave-assisted synthesis of porous aggregates of CuS nanoparticles for sunlight photocatalysis. *ACS Omega* **2019**, *4*, 4825–4831. [CrossRef]
5. Roy, P.; Srivastava, S.K. Nanostructured copper sulfides: synthesis, properties and applications. *CrystEngComm* **2015**, *17*, 7801–7815. [CrossRef]
6. Tanveer, M.; Cao, C.; Aslam, I.; Ali, Z.; Idrees, F.; Khan, W.S.; Tahir, M.; Khalid, S.; Nabi, G.; Mahmood, A. Synthesis of CuS flowers exhibiting versatile photo-catalyst response. *New J. Chem.* **2015**, *39*, 1459–1468. [CrossRef]

7. Wang, L. Synthetic methods of CuS nanoparticles and their applications for imaging and cancer therapy. *RSC Adv.* **2016**, *6*, 82596–82615. [CrossRef]

8. He, W.; Jia, H.; Li, X.; Lei, Y.; Li, J.; Zhao, H.; Mi, L.; Zhang, L.; Zheng, Z. Understanding the formation of CuS concave superstructures with peroxidase-like activity. *Nanoscale* **2012**, *4*, 3501–3506. [CrossRef]

9. Cheng, Z.; Wang, S.; Wang, Q.; Geng, B. A facile solution chemical route to self-assembly of CuS ball-flowers and their application as an efficient photocatalyst. *CrystEngComm* **2010**, *12*, 144–149. [CrossRef]

10. Xu, C.; Wang, L.; Zou, D.; Ying, T. Ionic liquid-assisted synthesis of hierarchical CuS nanostructures at room temperature. *Mater. Lett.* **2008**, *62*, 3181–3184. [CrossRef]

11. Mezgebe, M.M.; Ju, A.; Wei, G.; Macharia, D.K.; Guang, S.; Xu, H. Structure based optical properties and catalytic activities of hydrothermally prepared CuS nanostructures. *Nanotechnology* **2019**, *30*, 105704. [CrossRef]

12. Nafees, M.; Ali, S.; Rasheed, K.; Idrees, S. The novel and economical way to synthesize CuS nanomaterial of different morphologies by aqueous medium employing microwaves irradiation. *Appl. Nanosci.* **2012**, *2*, 157–162. [CrossRef]

13. Yao, K.; Zhao, C.; Sun, N.; Lu, W.; Zhang, Y.; Wang, H.; Wang, J. Freestanding CuS nanowalls: Ionic liquids assisted synthesis and prominent catalytic decomposition performance for ammonium perchlorate. *CrystEngComm* **2017**, *19*, 5048–5057. [CrossRef]

14. Kim, Y.; Heyne, B.; Abouserie, A.; Pries, C.; Ippen, C.; Günter, C.; Taubert, A.; Wedel, A. CuS nanoplates from ionic liquid precursors—application in organic photovoltaic cells. *J. Chem. Phys.* **2018**, *148*, 193818. [CrossRef] [PubMed]

15. Estrada, A.C.; Silva, F.M.; Soares, S.F.; Coutinho, J.A.P.; Trindade, T. An ionic liquid route to prepare copper sulfide nanocrystals aiming photocatalytic applications. *RSC Adv.* **2016**, *6*, 34521–34528. [CrossRef]

16. Yao, K.; Lu, W.; Li, X.; Wang, J. Ionic liquids-modulated two-phase thermal synthesis of three-dimensional CuS nanostructures. *J. Solid State Chem.* **2012**, *196*, 557–564. [CrossRef]

17. Lei, Z.; Chen, B.; Koo, Y.M.; MacFarlane, D.R. Introduction: ionic liquids. *Chem. Rev.* **2017**, *117*, 6633–6635. [CrossRef]

18. Yavir, K.; Marcinkowski, Ł.; Marcinkowska, R.; Namieśnik, J.; Kloskowski, A. Analytical applications and physicochemical properties of ionic liquid-based hybrid materials: a review. *Anal. Chim. Acta* **2019**, *1054*, 1–16. [CrossRef]

19. Du, Y.; Yin, Z.; Zhu, J.; Huang, X.; Wu, X.J.; Zeng, Z.; Yan, Q.; Zhang, H. A general method for the large-scale synthesis of uniform ultrathin metal sulfide nanocrystals. *Nat. Commun.* **2012**, *3*, 1177. [CrossRef]

20. Meng, X.; Tian, G.; Chen, Y.; Zhai, R.; Zhou, J.; Shi, Y.; Cao, X.; Zhou, W.; Fu, H. Hierarchical CuS hollow nanospheres and their structure-enhanced visible light photocatalytic properties. *CrystEngComm* **2013**, *15*, 5144–5149. [CrossRef]

21. Qian, J.; Zhao, Z.; Shen, Z.; Zhang, G.; Peng, Z.; Fu, X. A large scale of CuS nano-networks: catalyst-free morphologically controllable growth and their application as efficient photocatalysts. *J. Mater. Res.* **2015**, *30*, 3746–3756. [CrossRef]

22. Sreelekha, N.; Subramanyam, K.; Reddy, D.A.; Murali, G.; Varma, K.R.; Vijayalakshmi, R.P. Efficient photocatalytic degradation of rhodamine-B by Fe doped CuS diluted magnetic semiconductor nanoparticles under the simulated sunlight irradiation. *Solid State Sci.* **2016**, *62*, 71–81. [CrossRef]

23. Hu, H.; Wang, J.; Deng, C.; Niu, C.; Le, H. Microwave-assisted controllable synthesis of hierarchical CuS nanospheres displaying fast and efficient photocatalytic activities. *J. Mater. Sci.* **2018**, *53*, 14250–14261. [CrossRef]

24. Al-Kahtani, A.A. Photocatalytic degradation of rhodamine B dye in wastewater using gelatin/CuS/PVA nanocomposites under solar light irradiation. *J. Biomater. Nanobiotechnol.* **2017**, *8*, 66–82. [CrossRef]

25. Peng, S.; He, Y.; Er, M.; Sheng, Y.; Gu, Y.; Chen, H. Biocompatible CuS-based nanoplatforms for efficient photothermal therapy and chemotherapy in vivo. *Biomater. Sci.* **2017**, *5*, 475–484. [CrossRef] [PubMed]

26. Huang, J.; Zhou, J.; Zhuang, J.; Gao, H.; Huang, D.; Wang, L.; Wu, W.; Li, Q.; Yang, D.P.; Han, M.Y. Strong near-infrared absorbing and biocompatible CuS nanoparticles for rapid and efficient photothermal ablation of gram-positive and -negative bacteria. *ACS Appl. Mater. Inter.* **2017**, *9*, 36606–36614. [CrossRef]

27. Nikinmaa, M. What causes aquatic contamination? In *An Introduction to Aquatic Toxicology*, 1st ed.; Nikinmaa, M., Ed.; Academic Press: Cambridge, MA, USA, 2014; pp. 19–39.
28. Štengl, V.; Králová, D. TiO$_2$/ZnS/CdS nanocomposite for hydrogen evolution and orange II dye degradation. *Int. J. Photoenergy* **2011**, *2011*, 1–14. [CrossRef]

Sample Availability: Samples of the compounds are not available from the authors.

Article

Eco-Friendly Extraction and Characterisation of Nutraceuticals from Olive Leaves

Cinzia Benincasa [1,*] , Ilaria Santoro [2], Monica Nardi [3] , Alfredo Cassano [4] and Giovanni Sindona [2]

[1] CREA Research Centre for Olive, Citrus and TreeFruit, C.da Li Rocchi, I-87036 Rende, Cosenza, Italy
[2] Dipartimento di Chimica, Università della Calabria, Cubo 12C, I-87036 Rende, Cosenza, Italy; ilaria.santoro@unical.it (I.S.); giovanni.sindona@unical.it (G.S.)
[3] Dipartimento di Scienze della Salute, Università Magna Græcia, Viale Europa, I-88100 Germaneto, Cosenza, Italy; monica.nardi@unicz.it
[4] Institute on Membrane Technology, ITM-CNR, c/o University of Calabria, via P. Bucci, 17/C, I-87036 Rende, Cosenza, Italy; a.cassano@itm.cnr.it
* Correspondence: cinzia.benincasa@crea.gov.it; Tel.: +39-0984-4052209

Received: 5 September 2019; Accepted: 25 September 2019; Published: 25 September 2019

Abstract: Olive tree (*Olea europaea* L.) leaf, a waste by-product of the olive oil industry, is an inexpensive and abundant source of biophenols of great interest for various industrial applications in the food supplement, cosmetic, and pharmaceutical industries. In this work, the aqueous extraction of high-added value compounds from olive leaves by using microfiltered (MF), ultrapure (U), and osmosis-treated (O) water was investigated. The extraction of target compounds, including oleuropein (Olp), hydroxytyrosol (HyTyr), tyrosol (Tyr), verbascoside (Ver), lutein (Lut), and rutin (Rut), was significantly affected by the characteristics of the water used. Indeed, according to the results of liquid chromatography tandem mass spectrometry, the extracting power of microfiltered water towards rutin resulted very poor, while a moderate extraction was observed for oleuropein, verbascoside, and lutein. On the other hand, high concentrations of hydroxytyrosol were detected in the aqueous extracts produced with microfiltered water. The extraction power of ultrapure and osmosis-treated water proved to be very similar for the bio-active compounds oleuropein, verbascoside, lutein, and rutin. The results clearly provide evidence of the possibility of devising new eco-friendly strategies based on the use of green solvents which can be applied to recover bioactive compounds from olive leaves.

Keywords: green chemistry; olive leaves; natural products; tandem mass spectrometry

---

1. Introduction

In recent years, fundamental research has focused on using resources found in the environment for the protection of people's well-being [1]. Plant materials are widely used to maintain human health. Traditional medications, food supplements, and functional foods typically contain antioxidant compounds which may inhibit or decrease the rate of oxidation of other molecules by preventing the initiation and/or propagation of the chain reaction of free radicals [2]. Considering that plant materials are extremely complex matrices comprised of many components that can interfere with good separation, classic extraction procedures often involve different steps and the use of unsustainable solvents. Furthermore, the starting matrix from which these compounds are extracted is not always a waste product in the biomass processing industry. The development of green and environmentally friendly extraction methods of natural products is a hot research topic in the area of chemistry and technology [3]. In several experimental studies, phenols have demonstrated a wide spectrum of pharmacological activities beyond their antioxidant properties [4–6]. A potential source of these compounds is found in olive leaves: they are grouped with regard to major molecular characteristics as simple phenols and acids, lignans, and flavonoids [7] including flavones (luteolin-7-glucoside, apigenin-7-glucoside, diosmetin-7-glucoside, luteolin, and diosmetin), flavonols (rutin), flavan-3-ols (catechin), substituted phenols (tyrosol, hydroxytyrosol, vanillin, vanillic acid, and caffeic acid) [8,9], oleuropein, and other secoiridoids [10,11]. The latter are exclusive to the Oleaceae family. In fact, secoiridoids and other derivatives are the principal compounds of olive leaves [12], among which a major compound that is frequently reported is oleuropein. Flavonoids may occur in appreciable amounts [13] while simple phenols and acids are present in lower amounts.

The recovery of bioactive molecules from plant extracts, in view of the accepted use of these natural compounds as nutraceuticals [14–20], has gained a grown interest in the last decades. Nutraceuticals are commercially available and in great demand [21,22] as they possess the special role of preventing or even supporting medical therapies [23–25]. A very large number of studies describe the recovery of phenols from plant tissues [9–11,26,27], but all known methods applied to the extraction of phenols from leaves are based on the use of solvents, supercritical fluids, and classical analytical techniques using maceration assisted by liquid solvents [28]. Other methods can reduce solvent consumption or can use green solvents representing an environmental and economical alternative [15,19,29]. In recent years, concerns about the environmental impact have emerged as an issue of priority in society. New aspects related to the use of agro-industrial residues as by-products for further exploitation of high-value products are increasingly gaining interest, and their recovery may be economically attractive. Advances in biotechnology potentially offer opportunities for economic utilization of plant food residues such as grape and olive pomace, leaves, barks, roots, etc. The idea of turning "waste to wealth" by means of industrial food residues can considerably contribute to sustainable development. The by-products of the olive oil industry are an extraordinary source of bioactive phenolic compounds [19]. In this context, the concept of "Green Chemistry" has great importance in industrial processes to reduce or eliminate the use and generation of hazardous substances and was developed in principle to guide the chemists in their search towards greenness [30]. In the last decade, a new generation of green solvents and green methodologies to be used both in synthetic transformations and in extraction processes has been developed [31–33].

The use of water as solvent has attracted much interest in recent years [34,35]. Water features many benefits: water itself is not expensive, it can potentially improve reactivity and selectivity and enable the recycling of the catalyst [36–38], and it can allow mild reaction conditions in the use of protecting groups [39,40] and in the synthesis of bio-active compounds [41,42]. The study chemistry in water has also been an interesting way to gain insights into the biosynthesis and extraction of natural products [14,43,44].

To the best of our knowledge, the water-based extraction of phenolic compounds from olive leaves has been poorly investigated. Goldsmith et al. [45] used the response surface methodology (RSM) approach in order to identify the best possible combination of temperature, extraction time, and sample-to-solvent ratio for the aqueous extraction of phenolic compounds from olive leaves. The optimal conditions were proposed to be at 90 °C for 70 min at a sample-to-solvent ratio of 1:60 g/mL.

Ansari et al. [46] developed a green and inexpensive water-based procedure to extract oleuropein from olive leaf samples. The experimental results revealed that deionised water adjusted to pH 3 at 60 °C for 4 h had the highest extraction efficiency.

In this work we aimed to evaluate the aqueous extraction of bio-active compounds from whole and chopped olive leaves, such as oleuropein (Olp), hydroxytyrosol (HyTyr), tyrosol (Tyr), verbascoside (Ver), lutein (Lut), and rutin (Rut), by using ultrapure, microfiltered and osmosis-treated water. Tandem mass spectrometry analyses, extensively used in the field of structure evaluation of natural products, were performed to fully characterize the phenolic compounds of olive leaves.

2. Results and Discussion

2.1. Concentration and Trend of Bio-Active Compounds in Aqueous Extracts of Chopped Olive Leaves

Experimental results obtained from the aqueous extraction of chopped olive leaves are graphically summarized in Figure 1.

Water properties are very important for the extraction of Olp in chopped leaves: in particular, ultrapure (U) water appears to stimulate the migration process of Olp from the leaves to the solution more than micro-filtered (MF) and osmosis-treated (O) water. In fact, the extracting power of O and MF water was three and five times less than the extracting power of U water, respectively. In addition, for each type of water used for the extraction, the Olp concentration was affected by the time of infusion of the leaves: it reached its maximum concentration in the first day of infusion (753, 268, and 170 mg/kg in U, O, and MF water, respectively) and decreased up to reach a constant value after six days of infusion. The recovery of Olp resulted much higher if compared with some data reported in literature. Indeed, the use of water alone did not result in any detectable signal for Olp yield quantification as reported by Cifà et al. [47] where different pH of the water, times, and temperatures were tested. The authors, contrarily to us, finally established that water was not a good solvent to extract Olp from olive leaves. The Olp concentration in MF, O, and U water resulted also much higher than that reported by Ghomariet al. [48] when cold distilled water was used for the extraction (about 100 mg/kg). On the other hand, higher Olp concentrations were detected by using distilled water at 60 °C (19.3 ± 0.99 mg/g) and distilled water at 60 °C and pH 3 (23.36 ± 0.91 mg/g) [48]. Similarly, Ansari et al. [40] reported that distilled water at 60 °C and pH 3 for 4 h could allow the extraction of a large amount of Olp. In addition, Malik and Bradford [49] reported that although extraction in 80% methanol is the most effective method for olive leaf polyphenols, boiling of dried leaves was also a very efficient method for extracting Olp and Ver that gave 96% and 94% recoveries of these compounds, respectively, when compared with the methanol extract.

A prevalence of HyTyr in comparison with other bioactive compounds was detected in all aqueous extracts. These results are consistent with those reported by Herrero et al. [44] who showed that HyTyr was the main phenolic component on the water pressurized liquid extraction olive leaves extracts when water is used as extracting agent. On the other hand, oleuropein was the main component in the extracts obtained with ethanol. However, the extracting power of O and MF water resulted quite similar and almost ten times greater than the extracting power of U water (768, 736 and 56 mg/kg, respectively on the first day of infusion). HyTyr concentration slowly increased throughout the experimental period to reach high values at the end of the process (1139, 1008, and 189 mg/kg in O, MF, and U water, respectively). The recovery of HyTyr by using cold distilled water according to Ghomari et al. [48] was 200 mg/kg corresponding to our value at the end of the process with U water. The release of Tyr was, instead, very poor during all the experimental period and none of the three waters produced a significant difference (3.6, 2.2, and 1 mg/kg in MF, U, and O water, respectively, on the tenth day of infusion). These results were in agreement with those reported in the literature [48].

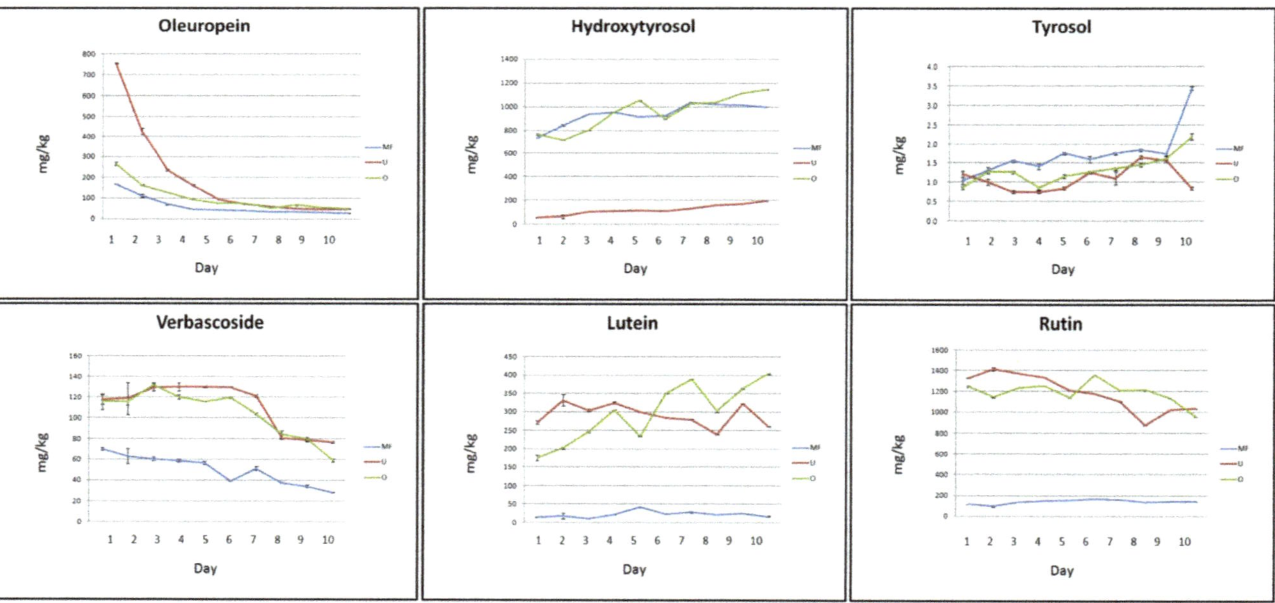

Figure 1. Trend of the concentrations of bio-active compounds in aqueous extracts of chopped olive leaves. Three types of water were tested: ultrapure (U), microfiltered (MF), and osmosis-treated (O) water. Data are expressed as the means ± standard error of the mean (SEM) of three independent observations. Statistical results from one-way analysis of variance (ANOVA) followed by Tukey-test are provided in Table S1 in Supporting Information.

U and O water showed a similar extracting power towards Ver, almost twice when compared to the extracting power of MF water (117 mg/kg in U and O water, 68 mg/kg in MF water, on the first day of infusion). The concentration of Ver reached its highest value on the sixth day of infusion in U water and on the third day of infusion in O water (133 and 134 mg/kg, respectively). Ver concentration decreased over the time to almost half its value by the end of the process (77, 57, and 27 mg/kg in U, O, and MF water, respectively). Ver resulted undetectable in the extraction by maceration in cold distilled water, as reported by Ghomari et al. [48]. A similar behaviour was observed for Lut: the migration process of this flavone from the leaves to the water was bland in MF water (15 and 18 mg/kg were the value registered from the start to the end of the infusion process) and more vigorous in U and O water (277and 178 mg/kg, respectively). The value of its concentration was higher on the sixth day of infusion in U water and on the last day of infusion in O water (356 and 400 mg/kg, respectively). The recovery of this flavone in olive leaves extracts obtained by maceration with distilled cold water was ten times lower (about 30 mg/kg) [48].

The release of Rut resulted very low in MF water during all the experimental process (between 91 and 175 mg/kg). On the other hand, the extracting power of U and O water was very incisive from the first day of infusion (1331 and 1244 mg/kg, respectively). Moreover, the concentration of Rut was not affected by the infusion time. The amount of rutin extracted in this study was much higher than that recorded by Ghomari et al. [48] who reported concentrations between 200 and 500 mg/kg.

The overall results indicated that with the exception of HyTyr and Tyr, the use of O and U water produced higher extraction efficiency of bioactive compounds, probably due to the absence or reduction of salt compounds which can affect the extraction process.

2.2. Concentration and Trend of Bio-Active Compounds in Aqueous Extracts of Whole Olive Leaves

As expected, the results obtained from the analysis of whole leaf infusions provided a concentration of bio-active compounds much lower than in the infusions discussed above. Figure 2 shows the trend of the concentrations of the bio-active compounds recorded in the whole olive leaves during the extraction process.

The migration of some bio-active compounds from whole leaves to the solution was not affected by the quality of the water. In particular, the extracting power of MF, U, and O water towards Olp was similar both at the beginning (86 mg/kg for MF and O water, and 81 mg/kg for U water) and at the end of the process (41 mg/kg for MF and O water, and 27 mg/kg for U water). Similarly, the extracting power of selected waters towards HyTyr resulted to be very similar at the beginning of the process (441, 467, and 512 mg/kg in MF, U, and O water, respectively, on the first day of infusion). HyTyr concentration reached the highest value on the sixth day of infusion in MF water (1152 mg/kg), on the seventh day of infusion in O water (1128 mg/kg), and on the eighth day of infusion in U water (1313 mg/kg).

As for chopped olive leaves, a prevalence of HyTyr in comparison with other bioactive compounds was detected in all aqueous extracts. These results are in agreement with those reported by Herrero et al. [50]. On the other hand, the release of Tyr was, again, very poor or null during all the experimental period, regardless of the type of water employed for the maceration process (3.4, 0.1, and 0.2 mg/kg in MF, U, and O water, respectively, on the tenth day of infusion).

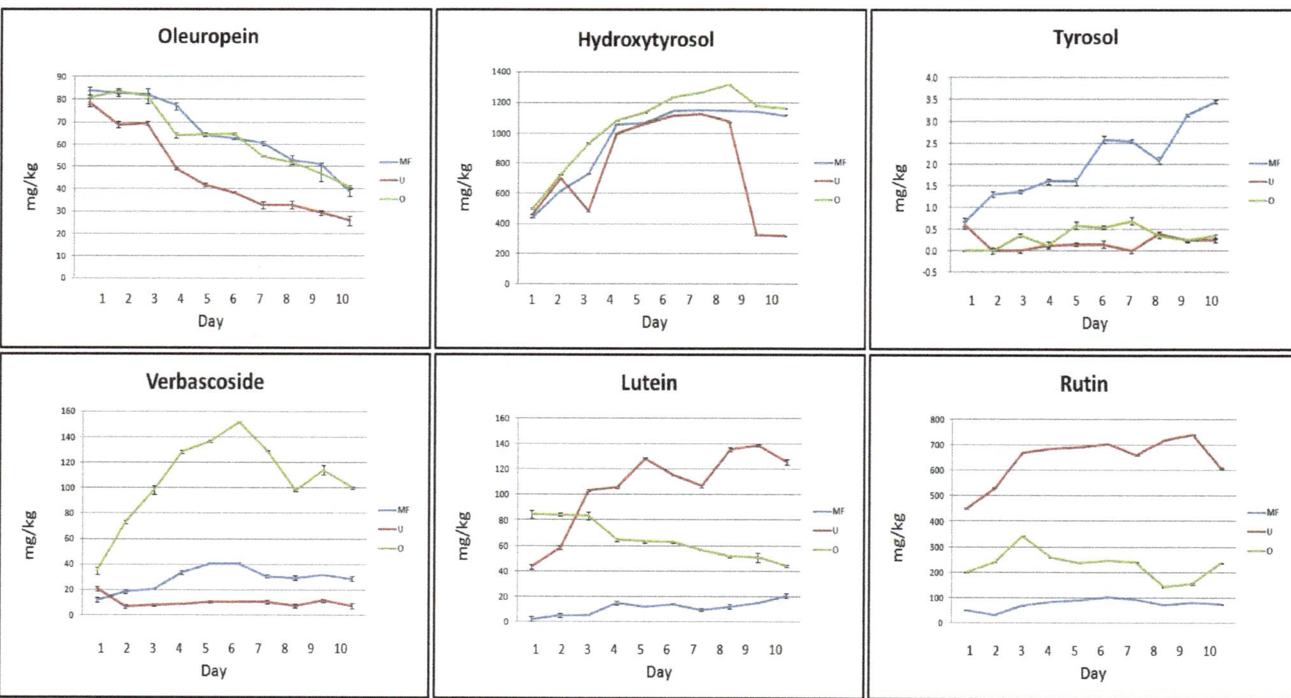

Figure 2. Trend of the concentrations of bio-active compounds in aqueous extracts of whole olive leaves. Three types of water were tested: ultrapure(U), microfiltered (MF), and osmosis-treated (O) water. Data are expressed as the means ± SEM of three independent observations. Statistical results from one-way analysis of variance (ANOVA) followed by Tukey-test are provided in Table S1 in Supporting Information.

The recovery of Ver, Lut, and Rut resulted to be influenced more greatly by the type of water. The concentration of Ver reached its highest value on the first day of infusion in U water (21 mg/kg) and on the sixth day of infusion in MF and O water (39 and 148 mg/kg, respectively). The migration process of Lut from the leaves to the water resulted very bland in MF water (3 and 19 mg/kg were the values registered from the start to the end of the infusion process) and gentle in U and O water (45 and 86 mg/kg, respectively). The concentration of the flavone constantly increased in U water, but decreased in O water (127 and 41 mg/kg, respectively). Similarly, the release of Rut resulted very low in MF water during the whole extraction process (between 34 and 112 mg/kg). The extracting power of U and O water, in contrast, gave some results from the first day of infusion (453 and 203 mg/kg, respectively). However, the concentration of the flavonol remained constant, or slightly increased, by the end of the process (610 and 239 mg/kg in U and O water, respectively).

The experimental results clearly indicated that the infusions of chopped leaves were richer in bio-active compounds in comparison to those of whole leaves. This could be explained by considering that the integrity of cell membranes of whole leaves does not promote the release of these bio-active compounds in water [51]. Furthermore, the glycosidic molecules such as Olp and Ver, even if extracted in important concentrations from the first day of infusion, were less stable over time compared to the other bio-active compounds analysed. This behaviour is probably due to the action of deglycosylation enzymes belonging to the leaves and released during the maceration process [52]. On the other hand, the migration process of HyTyr and Lut from the leaves to the solution has always shown to be increasing over time, while the extraction of Rut took place immediately and remained constant throughout the experimentation period.

The characteristics of the water used for the extraction processes were very important for the success of the extraction itself. In fact, for the flavonol Rut the extraction power of the MF water was very poor, while it was bland for Olp, Ver, and Lut. A completely different situation was found for HyTyr which, in MF water, was recovered in high concentrations.

The extraction power of U and O water proved to be very similar for the bio-active compounds Olp, Ver, Lut, and Rut. These waters have very similar chemical characteristics: U water contains by definition only H_2O, and H^+ and OH^- ions in equilibrium; osmosis-treated water is minimally mineralized. MF, instead, purified of chlorine and derivatives, dust and rust, preserves the mineral salts. These could in some way interfere with the process of migration of bio-active compounds (Rut), disturbing their accumulation in solution (Olp, Ver, and Lut), or even encouraging it (HyTyr).

2.3. Method Validation

For the determination of the best instrumental conditions, standard solutions of the selected bio-active compounds were introduced directly into the ion source of the mass spectrometer by direct infusion (FIA) at a flow rate of 10 μL/min. The mass spectrum obtained for Olp, a glycosylated secoiridoid, showed a pseudo molecular ion at *m/z* 539 and ionic fragments at *m/z* 307 and 275. These two characteristic ionic fragments originate from the ion at *m/z* 377 (a molecule resulting from the breakdown of the glycosidic bond of oleuropein which in ESI-MS experiments produces the ion at *m/z* 307 from the loss of a C_4H_6O fragment), and the ion at *m/z* 275 (derived from a rearrangement of other fragments). The mass spectrum obtained for HyTyr showed the deprotonated molecule [M − H]$^-$ at *m/z* 153 and the ionic fragment at *m/z* 123 due to the loss of a CH_2OH molecule. The mass spectrum obtained for Tyr did not produce any important fragments; therefore, the MRM measurements were conducted by scanning in both quadrupoles the deprotonated molecule [M − H]$^-$ at *m/z* 137. The mass spectrum obtained for Ver, a phenylpropanoid and an ester sugar of caffeic acid, showed an intense peak corresponding to the deprotonated molecule [M − H]$^-$ at *m/z* 623 and two characteristic ionic fragments at *m/z* 461 and 161. The loss of caffeic acid in fact produces an ion at *m/z* 461 and a neutral fragment, while the peak at *m/z* 161 results from a proton transfer and from the production of an anionic ketene. The mass spectrum obtained for Lut showed the deprotonated molecule [M − H]$^-$ at *m/z* 285 and fragments at *m/z* 133 and 151corresponding to retro Diels Alder fragmentation of flavone molecule.

The mass spectrum obtained for the flavonol Rut showed the deprotonated molecular ion at m/z 609 and a fragment at m/z 301 which is diagnostic of quercetin derivatives. The MRM transitions monitored for the assay of the bio-active compounds under investigation were, therefore, as follows: 539 → 307 and 539 → 275 for Olp; 153 → 123 for HyTyr; 137 → 137 for Tyr; 623 → 161 and 623 → 461 for Ver; 285 → 133 for Lut; 609 → 301 for Rut (Figure S1 in Supporting Information). Calibration curves constructed using a least-squares linear regression analysis were linear with correlation coefficients of 0.998 and 0.999 (Figure S2 in Supporting Information). Subsequently, to determine the best chromatographic conditions, the same standard solutions were injected into the HPLC-MS system through the chromatographic column (Figure 3). Spike solutions at 25 and 50 µg/mL gave good recoveries in a range between 87% and 109%, and satisfactory precision with relative standard deviation (RDS) in a range between 0.038% and 0.207% (Table 1). All data were obtained from three independent injections.

Table 1. Results from recovery tests of spike solutions at 25 and 50 µg/mL. Data are expressed as the means ± relative standard deviation (RSD) of three independent observations. Olp: oleuropein; HyTyr: hydroxytyrosol; Tyr: tyrosol; Ver: verbascoside; Lut: lutein; Rut: rutin.

	Spiked Solution (25 µg/mL)		Spiked Solution (50 µg/mL)	
Analyte	**Found** mean ± **RSD**	**Recovery** %	**Found** mean ± **RSD**	**Recovery** %
Olp	23.333 ± 0.108	93	50.667 ± 0.063	99
HyTyr	25.010 ± 0.080	100	52.333 ± 0.058	104
Tyr	24.333 ± 0.207	97	49.667 ± 0.081	99
Verb	24.667 ± 0.062	99	54.667 ± 0.038	109
Lut	21.667 ± 0.133	87	49.667 ± 0.111	99
Rut	22.512 ± 0.092	90	51.667 ± 0.095	103

3. Materials and Methods

3.1. Sample Collection and Preparation

Olive leaves from Coratina cultivar were collected from plants belonging to the olive grove of CREA-Research Centre for Olive, Citrus and TreeFruit, located in Rende (CS) and placed at 204 meters a.s.l. (39°22′17.681″N, 16°13′58.342″E).

Immediately after the harvest, plant tissues were processed for further experiments. Both whole and leaves roughly chopped by hands were considered in order to evaluate the best conditions for the extractions. In particular, 6 g of sample were placed in stoppered flasks containing 50 mL of water and kept at room temperature. The maceration processes were monitored by collecting each day, for 10 days, aliquots of the solutions and by performing high performance liquid chromatography tandem mass spectrometric analyses (LC-MS/MS). The maceration process of olive leaves was carried out by applying low-cost and eco-friendly methods where no solvents and supercritical fluids were used.

Three types of water were tested: ultrapure (U), microfiltered (MF), and osmosis-treated (O) water. U water was produced through a Milli-Q plus water purification system (Millipore, Bedford, MA, USA) which includes a four-cartridge purification pack (Qpak), using deionized water as feed water. This module includes activated-carbon and ion-exchange resins housed within four polypropylene cylinders. A final filter (pore size 0.22 µm), within polycarbonate housing, was part of the dispensing point of this unit. The produced water was characterized by an electrical conductivity of 0.055 µS/cm. MF and O water were obtained from tap water. MF was performed by using a bench laboratory plant consisting of a stainless steel feed tank, a magnetic drive gear pump and a stainless-steel cell able to accommodate a flat-sheet membrane with a surface area of 38.46 cm^2. The cell was equipped with a polyvinylidene fluoride membrane (MV020 T) with a nominal pore size of 0.2 µm, supplied by Microdyn-Nadir GmbH (Wiesbaden, Germany). Transmembrane pressure (TMP) was measured by two manometers allocated before and after the membrane cell and regulated by a pressure control valve on the concentrate outlet. Crossflow velocity (CFV) was controlled by a digital flowmeter. Temperature was controlled by using a cooling system fed with tap water and monitored by a digital thermometer inserted in the feed tank. Tap water was microfiltered at a TMP of 0.5 bar and an operating temperature of 25 ± 2 °C, producing a permeate water stream with an electrical conductivity of about 810 μS/cm. Reverse osmosis (RO) was performed by using a RO laboratory bench plant consisting of a control panel, a cylindrical jacketed feed tank (with a capacity of 5 L) constructed from stainless steel (SS 316), a feed plunger pump with belt drive (Cat Pumps, Milano, Italy, Model 3CP1221), two pressure gauges (Wika Instrument, Lawrenceville, GA, USA) (max pressure 100 bar, absolute error 1 bar), a digital flow meter (SM6000, ifm electronic gmbh, Essen, Germany), a thermometer placed inside the feed tank, and a cylindrical housing able to accommodate a 11.74 × 1.75-inchspiral-wound membrane module. The adjustment of operating pressure and feed flow rate was done by simultaneously pump rotation control through a frequency inverter and a needle valve. The operating temperature was controlled by circulating a coolant (cold water) through the tank jacket. The plant was equipped with a thin-film polyamide membrane module (SC1812-34D), supplied by GE Water & Process Technologies (Hopkins, MN, USA), having a NaCl rejection of 99.5% and a membrane surface area of 0.32 m^2. The system was operated at a TMP of 6 bar and an operating temperature of 25 ± 2 °C producing a permeate water stream with an electrical conductivity of 42.7 μS/cm.

3.2. Quantitative Analysis

Standards of oleuropein (Olp), tyrosol (Tyr), and 3-hydroxytyrosol (HyTyr) were purchased from Sigma–Aldrich (Sigma-Aldrich, St. Louis, MO, USA); verbascoside (Ver), lutein (Lut), and rutin (Rut) were purchased from Extrasynthese (Genay, France). All solvents were of LC/MS grade and purchased from VWR International. The assay of the bio-active analytes was achieved by external standard calibration: standard stock solutions were prepared by dissolving reference compounds in ethanol. Aliquots of these solutions were further diluted with water/0.1% formic acid to obtain calibration standards at concentrations between 1 and 200 μg/mL for Olp and HyTyr, 1 and 100μg/mL for Tyr, Lut, and Rut, and 1 and 150 μg/mL for Ver. The performance of the experiments was checked by recovery tests of spike solutions (Table 1).

3.3. High-Performance Liquid Chromatography/Tandem Mass Spectrometry

HPLC analysis were performed using an Agilent Technologies 1200 series liquid chromatography system equipped with G1379B degasser, G1312A pump, and G1329A autosampler. The chromatographic separation was achieved by injecting 10 μL of ethanol extract in an Eclipse XDB-C8-A column (5 μm particle size; 150 mm length; 4.6 mm internal diameter) (Agilent Technologies, Santa Clara, CA, USA) and a mobile phase consisted of an aqueous formic acid (0.1%) solution (A) and methanol (B). The separation of the analytes was achieved in 25 min at a flow rate of 350 μL/min. In particular, the elution profile was as follows: 0–10 min, 10–100% B (*v/v*); 10–12 min, 100% B; 12–20 min, 100–10% B (*v/v*); and 20–25 min, 100–10% B (*v/v*). The time for the column to re-equilibrated was 5 min. The mass spectrometer utilized for MS/MS analyses was an API 4000 Q-Trap (Applied Biosystem/MSD Sciex, Foster City, CA, USA). Each compound was monitored in negative ion mode using multiple reaction monitoring (MRM). The instrumental parameters, such as entrance potential (EP), declustering potential (DP), collision energy (CE), and collision exit potential (CXP) were, therefore, optimized for each transition monitored. Ionspray voltage (IS) was set at –4500 V; curtain gas at 20 psi; temperature at 400 °C; ion source gas(1) at 35 psi; ion source gas(2) at 45 psi; and collision gas thickness (CAD) medium. The HPLC-MS chromatograms of the bioactive compounds analysed in MRM mode are shown in Figure 3. In particular, the chromatograms refer to the analysis of the analytes of interest in the aqueous solutions of whole olive leaves after five days of maceration with: ultrapure water (A), microfiltered water (B) and osmosis-treated water (C). The retention times, expressed in minutes (min) are shown on the abscissas, whilst the intensity signals, expressed as counts per scan (cps), on the ordinates.

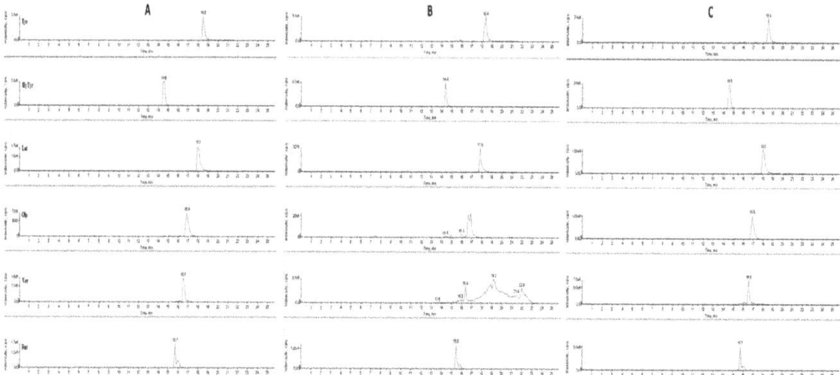

Figure 3. HPLC-MS chromatograms in multiple reactions monitoring MRM mode of bio-active compounds in the aqueous solutions of whole olive leaves after five days of maceration with: ultrapure water (**A**); microfiltered water (**B**) and osmosis-treated water (**C**).

3.4. Statistical Analysis

The results are expressed by mean ± standard error of the mean (SEM) from at least three independent experiments. For statistical comparisons, quantitative data were analysed by one-way analysis of variance (ANOVA) followed by Tukey-test according to the statistical program SigmaStat1 (Jandel Scientific, Chicago, IL, USA). As the *p*-values were less than 0.001, there is a strong evidence for significant differences between the aqueous extraction of bio-active compounds from whole and chopped olive leaves by using ultrapure, microfiltered and osmosis-treated water (Table S1 in Supporting Information).

4. Conclusions

The agro-food industries are focused on the use of bio-active compounds and nutraceuticals obtained from industrial waste products. Even though conventional extraction still is the main approach for obtaining bio-active compounds, this technology is not aligned with "green" and sustainable production, as it is very often accompanied by high expenditure and disposal of energy and toxic chemicals. The eco-friendly separation of bio-compounds from agro-industrial waste, supported by the use of an eco-sustainable and low-cost solvent such as water, is obviously attractive from both socio-environmental and economic points of view. Opening up new occasions for ecological approaches designed for bio-economy and circular economy models, the future food manufacturing can foresee better solutions for industrial production and applications. With the aim to increase production and process efficiency, reducing solvent and energy consumption and decreasing food waste by improving shelf life, may certainly have an important impact in changing industrial and academic practices. The previously discussed results clearly provide evidence for the possibility of devising new eco-friendly strategies that can be applied to recover important active compounds from olive leaves. In fact, from the sole use of water it was possible to obtain extracts of molecules of great health and pharmacological value by using a cheap and renewable source of natural product that is also an agricultural and industrial waste. In the procedure described, the use of waste material and the eco-sustainable and easily available solvents are proposed to contribute to sustainable development (Figure 4). The results that have been achieved were the followings: the infusions of chopped leaves were richer in bio-active compounds than infusions of whole leaves. This could be explained by considering that the integrity of cell membranes of whole leaves does not promote the release of these bio-active compounds in the water solutions. Furthermore, the glycosidic molecules such as Olp and Ver, even if extracted in important concentrations from the first day of infusion, were less stable over time compared to the other bio-active compounds analysed. This behaviour is probably due to the action of deglycosylation enzymes belonging to the leaves and released during the maceration process. On the other hand, the migration process of HyTyr and Lut from the leaves to the solution has always been shown to increase over time, while extraction of Rut took place immediately and remained constant throughout the experimentation period. The characteristics of the water used for the extraction processes were very important for the success of the extraction itself. In fact, for the flavonol Rut the extraction power of the MF water was null, while it was bland for Olp, Ver, and Lut. A completely different situation was found for HyTyr which, in MF water, was recovered in high concentrations. The extraction power of U and O water proved to be very similar for the bio-active compounds Olp, Ver, Lut, and Rut. These latter waters have very similar chemical characteristics: ultrapure water contains by definition only H_2O and H^+ and OH^- ions in equilibrium; osmosis-treated water is minimally mineralized. The MF water, purified of chlorine and derivatives, dust, and rust, preserves the mineral salts. These could in some way interfere with the process of migration of bio-active compounds (Rut), disturbing their accumulation in solution (Olp, Ver, and Lut), or even encouraging it (HyTyr).

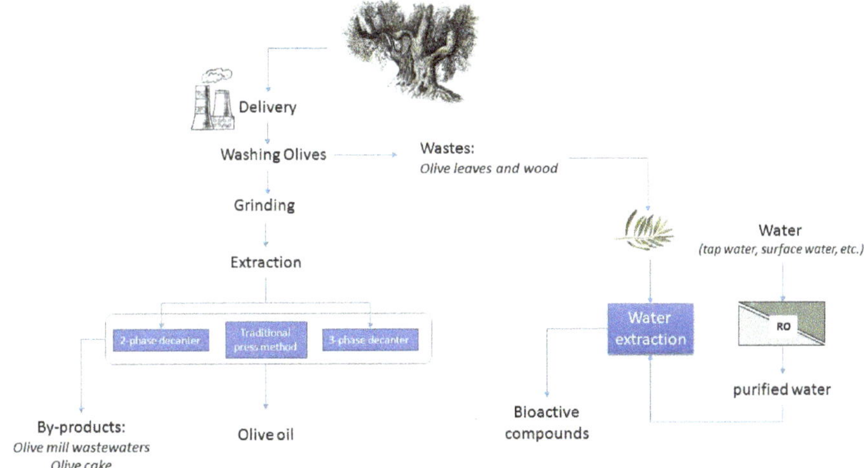

Figure 4. Scheme of olive oil production and aqueous extracts of highly nutritional and pharmacological value from olive leaves (RO, reverse osmosis).

Supplementary Materials: The following are available online, Figure S1: LC-MS/MS spectra of the main phenolic compounds analysed: oleuropein (Olp), hydroxytyrosol (HyTyr), tyrosol (Tyr), lutein (Lut), verbascoside (Ver), and rutin (Rut), showing the deprotonated molecular ion $[M - H]^-$ in the main fragments used in the MRM method; Figure S2: HPLC-MS/MS external calibration curves with equation and correlation coefficient R^2; Table S1: Statistical results from one-way analysis of variance (ANOVA) of the quantitative data of the selected bio-active compounds monitored by using HPLC-MRM methodology. The aqueous extracts of whole and chopped olive leaves were obtained by using three types of water: ultrapure(U), microfiltered (MF), and osmosis-treated (O) water. Highlighted boxes indicate a significant difference with a p-value less than 0.001.

Author Contributions: G.S. and C.B. designed the study and the experiments; A.C. provided water samples from membrane filtration; I.S. carried out the collection and the extraction of samples; C.B. analysed the water extracts by means of HPLC-MS/MS; C.B., M.N., and A.C. wrote, edited, and drafted the manuscript.

Funding: This work was supported by the Calabria region under the project APQ-RAC QUASIORA. Ilaria Santoro thanks the doctorate school of Translational Medicine for a fellowship. The authors are grateful to Mr. Massimiliano Pellegrino for providing technical help and support during all the research activities at CREA Research Centre for Olive, Citrus and Tree Fruit of Rende (CS).

Conflicts of Interest: The authors declare no conflict of interest.

References

1. Vlek, C.; Steg, L. Human behavior and environmental sustainability: Problems, driving forces, and research topics. *J. Soc. Issues* **2007**, *63*, 1–19. [CrossRef]

2. Valduga, A.T.; Goncalves, I.L.; Magri, E.; Delalibera Finzer, J.R. Chemistry, pharmacology and new trends in traditional functional and medicinal beverages. *Food Res. Int.* **2019**, *120*, 478–503. [CrossRef] [PubMed]

3. Chemat, F.; Vian, M.A.; Cravotto, G. Green extraction of natural products: Concept and principles. *Int. J. Mol. Sci.* **2012**, *13*, 8615–8627. [CrossRef] [PubMed]

4. Hassen, I.; Casabianca, H.; Hosni, K. Biological activities of the natural antioxidant oleuropein: Exceeding the expectation–A mini-review. *J. Funct. Food.* **2015**, *18*, 926–940. [CrossRef]

5. Lupinacci, S.; Toteda, G.; Vizza, D.; Perri, A.; Benincasa, C.; Mollica, A.; La Russa, A.; Gigliotti, P.; Leone, F.; Lofaro, D.; et al. Active compounds extracted from extra virgin olive oil counteract mesothelial-to-mesenchymal transition of peritoneal mesothelium cells exposed to conventional peritoneal dialysate: In vitro and in vivo evidences. *J. Nephrol.* **2016**, *30*, 841–850. [PubMed]

6. Farooqi, A.A.; Fayyaz, S.; Sanches Silva, A.; Sureda, A.; Nabavi, S.F.; Mocan, A.; Nabavi, S.F.; Mocan, A.; Nabavi, S.M.; Bishayee, A. Oleuropein and cancer chemo prevention: The link is hot. *Molecules* **2017**, *22*, 705. [CrossRef] [PubMed]

7. Tsimidou, M.Z.; Papoti, V.T. Bioactive Ingredients in olive leaves. In *Olives and Olive Oil in Health and Disease Prevention*; Preedy, V.R., Watson, R.R., Eds.; Academic Press: San Diego, CA, USA, 2010; pp. 349–356.

8. Konno, K.; Hirayama, C.; Yasui, H.; Nakamura, M. Enzymatic activation of oleuropein: A protein crosslinker used as a chemical defense in the privet tree. *Proc. Natl. Acad. Sci. USA* **1999**, *96*, 9159–9164. [CrossRef]

9. El, S.N.; Karakaya, S. Olive tree (*Oleaeuropaea*) leaves: Potential beneficial effects on human health. *Nutr. Rev.* **2009**, *67*, 632–638. [CrossRef]

10. Briante, R.; La Cara, F.; Tonziello, M.P.; Febbraio, F.; Nucci, R. Antioxidant activity of the main bioactive derivatives from oleuropein hydrolysis by hyperthermophilic β-glycosidase. *J. Agric. Food Chem.* **2001**, *49*, 3198–3203. [CrossRef]

11. Briante, R.; Patumi, M.; Limongelli, S.; Febbraio, F.; Vaccaro, C.; Di Salle, A.; La Cara, F.; Nucci, R. Changes in phenolic and enzymatic activities content during fruit ripening in two Italian cultivars of *Olea europaea* L. *Plant Sci.* **2002**, *162*, 791–798. [CrossRef]

12. Kontogianni, V.G.; Gerothanassis, I.P. Phenolic compounds and antioxidant activity of olive leaf extracts. *Nat. Prod. Res.* **2012**, *26*, 186–189. [CrossRef] [PubMed]

13. Savournin, C.; Baghdikian, B.; Elias, R.; Dargouth-Kesraoui, F.; Boukef, K.; Balansard, G. Rapid high-performance liquid chromatography analysis for the quantitative determination of oleuropein in *Olea europaea* leaves. *J. Agric. Food Chem.* **2001**, *49*, 618–621. [CrossRef] [PubMed]

14. Benincasa, C.; Perri, E.; Romano, E.; Santoro, I.; Sindona, G. Nutraceuticals from olive plain water extraction identification and assay by LC-ESI-MS/MS. *J. Anal. Bioanal. Tech.* **2015**, *6*, 1–8.

15. Didaskalou, C.; Buyuktiryaki, S.; Kecili, R.S.; Fonte, C.P.; Szekely, G. Valorisation of agricultural waste with an adsorption/nanofiltration hybrid process: From materials to sustainable process design. *Green Chem.* **2017**, *19*, 3116–3125. [CrossRef]

16. Procopio, A.; Alcaro, S.; Nardi, M.; Oliverio, M.; Ortuso, F.; Sacchetta, P.; Pieragostino, D.; Sindona, G. Synthesis, biological evaluation, and molecular modeling of oleuropein and its semisynthetic derivatives as cyclooxygenase inhibitors. *J. Agric. Food Chem.* **2009**, *57*, 11161–11167. [CrossRef] [PubMed]

17. Oliverio, M.; Costanzo, P.; Nardi, M.; Calandruccio, C.; Salerno, R.; Procopio, A. Tunable microwave-assisted method for the solvent-free and catalyst-free peracetylation of natural products. *Beilstein J. Org. Chem.* **2016**, *12*, 2222–2233. [CrossRef] [PubMed]

18. Nardi, M.; Bonacci, S.; Cariati, L.; Costanzo, P.; Oliverio, M.; Sindona, G.; Procopio, A. Synthesis and antioxidant evaluation of lipophilic oleuropein aglycone derivatives. *Food Funct.* **2017**, *8*, 4684–4692. [CrossRef] [PubMed]

19. Sahin, S.; Samli, R.; Birteksöz Tan, A.S.; Barba, F.J.; Chemat, F.; Cravotto, G.; Lorenzo, J.M. Solvent-free microwave-assisted extraction of polyphenols from olive tree leaves: Antioxidant and antimicrobial properties. *Molecules* **2017**, *22*, 1056. [CrossRef] [PubMed]

20. Costanzo, P.; Bonacci, S.; Cariati, L.; Nardi, M.; Oliverio, M.; Procopio, A. Simple and efficient sustainable semi-synthesis of oleacein [2-(3,4-hydroxyphenyl) ethyl (3*S*,4*E*)-4-formyl-3-(2-oxoethyl)hex-4-enoate] as potential additive for edible oils. *Food Chem.* **2018**, *245*, 410–414. [CrossRef]

21. Ciafardini, G.; Marsilio, V.; Lanza, B.; Pozzi, N. Hydrolysis of oleuropein by Lactobacillus plantarum strains associated with olive fermentation. *Appl. Envir. Micr.* **1994**, *60*, 4142–4147.

22. Angerosa, F.; D'Alessandro, N.; Cornara, F.; Mellerio, G. Characterization of phenolic and secoiridoid aglycones present in virgin olive oil by gas chromatography-chemical ionization mass spectrometry. *J. Chrom. A* **1996**, *736*, 195–203. [CrossRef]

23. Sindona, G. A marker of quality of olive oils: The expression of oleuropein. In *Olives and Olive Oil in Health and Disease Prevention*; Preedy, V.R., Watson, R.R., Eds.; Academic Press: San Diego, CA, USA, 2010; pp. 95–100.

24. Bulotta, S.; Celano, M.; Lepore, S.M.; Montalcini, T.; Pujia, A.; Russo, D. Beneficial effects of the olive oil phenolic components oleuropein and hydroxytyrosol: Focus on protection against cardiovascular and metabolic diseases. *J. Transl. Med.* **2014**, *12*, 1–9. [CrossRef] [PubMed]

25. Fabiani, R. Anti-cancer properties of olive oil secoiridoid phenols: A systematic review of in vivo studies. *Food Funct.* **2016**, *7*, 4145–4159. [CrossRef] [PubMed]

26. Ryan, D.; Antolovich, M.; Prenzler, P.; Robards, K.; Lavee, S. Bio transformations of phenolic compounds in *Olea europaea* L. *Sci. Hort.* **2002**, *92*, 147–176. [CrossRef]

27. Morello, J.R.; Motiva, M.J.; Tovar, M.J.; Romero, M.P. Changes in commercial olive oil (cv Arbequina) during storage with special emphasis on phenolic fraction. *Food Chem.* **2004**, *85*, 357–364. [CrossRef]

28. Bilgin, M.; Sahin, S. Effects of geographical origin and extraction methods on total phenolic yield of olive tree (*Oleaeuropaea*) leaves. *J. Taiwan Inst. Chem. Eng.* **2013**, *44*, 8–12. [CrossRef]

29. Chemat, F.; Fabiano-Tixier, A.S.; Vian, M.A.; Allaf, T.; Vorobiev, E. Solvent-free extraction. *Compr. Anal. Chem.* **2017**, *76*, 225–254.

30. Anastas, P.; Eghbali, N. Green chemistry: Principles and practice. *Chem.Soc. Rev.* **2010**, *39*, 301–312. [CrossRef]

31. Tommasi, E.; Cravotto, G.; Galletti, P.; Grillo, G.; Mazzotti, M.; Tacchini, M.; Tagliavini, E. Enhanced and selective lipid extraction from the microalga P. tricornutum by dimethyl carbonate and supercritical CO$_2$ using deep eutectic solvents and microwaves as pretreatment. *ACS Sustainable Chem. Eng.* **2017**, *5*, 8316–8322. [CrossRef]

32. Mourtzinos, I.; Anastasopoulou, E.; Athanasios, P.; Grigorakis, S.; Makris, D.; Biliaderis, C.G. Optimization of a green extraction method for the recovery of polyphenols from olive leaf using cyclodextrins and glycerin as co-solvents. *J. Food Sci. Technol.* **2016**, *53*, 3939–3947. [CrossRef]

33. Apostolakis, A.; Grigorakis, S.; Makris, D. Optimisation and comparative kinetics study of polyphenol extraction from olive leaves (*Olea europaea*) using heated water/glycerol mixtures. *Sep. Purif. Technnol.* **2014**, *128*, 89–95. [CrossRef]

34. Grieco, P.A. *Organic Synthesis in Water*; Blackie Academic & Professional: London, UK, 1998.

35. Marc-Olivier, S.; Chao-Jun, L. Green chemistry oriented organic synthesis in water. *Chem. Soc. Rev.* **2012**, *41*, 1415–1427.

36. Procopio, A.; Gaspari, M.; Nardi, M.; Oliverio, M.; Rosati, O. Highly efficient and versatile chemo selective addition of amines to epoxides in water catalyzed by erbium(III)triflate. *Tetrahedron Lett.* **2008**, *49*, 2289–2293. [CrossRef]

37. Procopio, A.; Cravotto, G.; Oliverio, M.; Costanzo, P.; Nardi, M.; Paonessa, R. An eco-sustainable erbium(III)-catalyzed method for formation/cleavage of O-tert-butoxy carbonates. *Green Chem.* **2011**, *13*, 436–443. [CrossRef]

38. Nardi, M.; Di Gioia, M.L.; Costanzo, P.; De Nino, A.; Maiuolo, L.; Oliverio, M.; Olivito, F.; Procopio, A. Selective acetylation of small biomolecules and their derivatives catalyzed by Er(OTf)$_3$. *Catalysts* **2017**, *7*, 269. [CrossRef]

39. Procopio, A.; Gaspari, M.; Nardi, M.; Oliverio, M.; Tagarelli, A.; Sindona, G. Simple and efficient MW-assisted cleavage of acetals and ketals in pure water. *Tetrahedron Lett.* **2007**, *48*, 8623–8627. [CrossRef]

40. Nardi, M.; Herrera Cano, N.; Costanzo, P.; Oliverio, M.; Sindona, G.; Procopio, A. Aqueous MW eco-friendly protocol for amino groupprotection. *RSC Adv.* **2015**, *5*, 18751–18760. [CrossRef]

41. Cano, N.H.; Uranga, J.G.; Nardi, M.; Procopio, A.; Wunderlin, D.A.; Santiago, A.N. Selective and eco-friendly procedures for the synthesis of benzimidazole derivatives. The role of the Er(OTf)$_3$ catalyst in the reaction selectivity. *Beilstein J. Org. Chem.* **2016**, *12*, 2410–2419. [CrossRef] [PubMed]

42. Nardi, M.; Costanzo, P.; De Nino, A.; Di Gioia, M.L.; Olivito, F.; Sindona, G.; Procopio, A. Water excellent solvent for the synthesis of bifunctionalizedcyclopentenones from furfural. *Green Chem.* **2017**, *19*, 5403–5411. [CrossRef]

43. Oliverio, M.; Nardi, M.; Cariati, L.; Vitale, E.; Bonacci, S.; Procopio, A. "On Water" MW-assisted synthesis of hydroxytyrosol fatty esters. *ACS Sustainable Chem. Eng.* **2016**, *4*, 661–665. [CrossRef]

44. Putnik, P.; Kovačević, D.B.; Jambrak, A.R.; Barba, F.J.; Cravotto, G.; Binello, A.; Lorenzo, J.M.; Shpigelman, A. Innovative "green" and novel strategies for the extraction of bioactive added value compounds from citrus wastes–a review. *Molecules* **2017**, *22*, 680. [CrossRef] [PubMed]

45. Goldsmith, C.D.; Vuong, Q.V.; Stathopoulos, C.E.; Roach, P.D.; Scarlett, C.J. Optimization of the aqueous extraction of phenolic compounds from olive leaves. *Antioxidants* **2014**, *3*, 700–712. [CrossRef] [PubMed]

46. Ansari, M.; Kazemipour, M.; Fathi, S. Development of a simple green extraction procedure and HPLC method for determination of oleuropein in olive leaf extract applied to a multi-source comparative study. *J. Iran Chem. Soc.* **2011**, *8*, 38–47. [CrossRef]

47. Cifá, D.; Skrt, M.; Pittia, P.; Di Mattia, C.; Ulrih, N.P. Enhanced yield of oleuropein from olive leaves using ultrasound-assisted extraction. *Food Sci. Nutr.* **2018**, *6*, 1128–1137. [CrossRef] [PubMed]

48. Ghomari, O.; Sounni, F.; Massaoudi, Y.; Ghanam, J.; DrissiKaitouni, L.B.; Merzouki, M.; Benlemlih, M. Phenolic profile (HPLC-UV) of olive leaves according to extraction procedure and assessment of antibacterial activity. *Biotechnol. Rep.* **2019**, *23*, e00347. [CrossRef] [PubMed]

49. Malik, N.S.A.; Bradford, J.M. Recovery and stability of oleuropein and other phenolic compounds during extraction and processing of olive (*Oleaeuropaea* L.) leaves. *J. Food Agric. Environ.* **2008**, *6*, 8–13.

50. Herrero, M.; Temirzoda, T.N.; Segura-Carretero, A.; Quirantes, R.; Plaza, M.; Ibañez, E. New possibilities for the valorization of olive oil by-products. *J. Chromatogr. A* **2011**, *1218*, 7511–7520. [CrossRef] [PubMed]

51. Xu, D.P.; Li, Y.; Meng, X.; Zhou, T.; Zhou, Y.; Zheng, J.; Zhang, J.J.; Li, H.B. Natural antioxidants in foods and medicinal plants: Extraction, assessment and resources. *Int. J. Mol. Sci.* **2017**, *18*, 96. [CrossRef]

52. Silva, S.; Gomes, L.; Leitao, F.; Coelho, A.V.; Boas, L.V. Phenolic compounds and antioxidant activity of *Olea europaea* L. fruits and leaves. *Food Sci. Technol. Int.* **2006**, *12*, 385–395. [CrossRef]

Sample Availability: Not available.

 *molecules*

Article

Protic Ionic Liquids as Efficient Solvents in Microwave-Assisted Extraction of Rhein and Emodin from *Rheum palmatum* L.

Yunchang Fan [1],* , Zeyu Niu [2], Chen Xu [1], Lei Yang [1] and Tuojie Yang [1]

1   College of Chemistry and Chemical Engineering, Henan Polytechnic University, Jiaozuo 454003, China
2   College of Food Science and Engineering, Central South University of Forestry and Technology, Changsha 410004, China
*   Correspondence: fanyunchang@hpu.edu.cn; Tel.: +86-391-398-6813

Academic Editors: Monica Nardi, Antonio Procopio and Maria Luisa Di Gioia
Received: 5 July 2019; Accepted: 29 July 2019; Published: 30 July 2019

Abstract: *Rheum palmatum* L. (*R. palmatum* L.) is a traditional Chinese herb and food, in which rhein and emodin are the main bioactive components. The extraction of the two compounds from *R. palmatum* L. is, thus, of great importance. In this work, protic ionic liquids (PILs) were applied in the microwave-assisted extraction (MAE) of rhein and emodin from *R. palmatum* L., which avoids the toxicity of organic solvents. The results of the present study indicate that PILs possessing higher polarity exhibit higher extraction ability due to their stronger absorption ability for microwave irradiation. Compared with conventional solvents, such as methanol, trichloromethane, and deep eutectic solvents (DESs), the PIL, 1-butyl-3-himidazolium methanesulfonate ([BHim]MeSO$_3$) reported herein is more efficient. The selected extraction conditions of liquid–solid ratio, microwave irradiation time, microwave irradiation power, and PIL concentration were 40 g·g^{-1}, 50 s, 280 W, and 80%, respectively. Under the selected conditions, the extraction yields of rhein and emodin were 7.8 and 4.0 mg·g^{-1}, respectively. These results suggest that PILs are efficient extraction solvents for the separation of active components from natural products.

Keywords: microwave-assisted extraction; protic ionic liquids (PILs); polarity; *Rheum palmatum* L.

1. Introduction

Rheum palmatum L., a well-known traditional Chinese herb and food, is used to treat various diseases and adverse conditions, such as high fever, intense sweating, constipation, and abdominal pain [1,2]. Rhein and emodin, as the major components of *R. palmatum* L., were investigated in depth and are widely used in pharmaceuticals, functional foods, and natural yellow dyes because of their excellent bioactivities [3,4]. Consequently, an efficient green extraction method should be developed to extract rhein and emodin from *R. palmatum* L. The traditional extraction of rhein and emodin from *R. palmatum* L. is usually carried out using many time-consuming and inefficient conventional extraction techniques, such as marinated extraction (ME), heat reflux extraction (HRE), and Soxhlet extraction [5]. Modern extraction techniques such as microwave-assisted extraction (MAE) provide an efficient alternative with the advantages of higher extraction efficiency, faster extraction processes, lower solvent consumption etc. [6]. However, the remarkable deficiency of MAE is that the abundant volatile organic solvents used in the extraction process have detrimental effects on the environment and human health. Ionic liquids (ILs), constituting organic cations and inorganic anions with unique properties, such as a low melting temperature, high thermal stability, wide liquid phase range, and low vapor pressure, were used in the extraction of natural products [7–9]; however, 1,3-dialkylimidazolium-based ILs used in previous extraction processes entail a high cost [10–12]. Therefore, the use of green and economic solvents instead of hazardous solvents is one of the most important considerations during extraction [13,14]. In this context, Procopio and Nardi's group used the MAE method with an eco-sustainable solvent, water, as the extraction solvent to extract oleuropein from olive leaves [15,16]. Another interesting work from the same research group also reported the extraction of demethyloleuropein from ripened drupes of the Leccino cultivar using methanol maceration at room temperature [17].

Additionally, Wu et al. proposed an ultrasound-assisted extraction (UAE) method for the extraction of total anthraquinones from *R. palmatum* L. using natural deep eutectic solvents (DESs) [18]. The reported DES, consisting of lactic acid, glucose, and water (LGH), was highly efficient in extracting anthraquinones from *R. palmatum* L., and it possesses many noteworthy advantages, such as low cost, easy preparation, biodegradability, pharmaceutically acceptable toxicity, and sustainability [19,20]. Inspired by their work, we tried to develop a fast, highly effective, environmentally benign, and low-cost method for the extraction of *R. palmatum* L. through changing DESs into protic ionic liquids (PILs) and using the MAE method instead of UAE. The MAE technique was adopted due to its unique properties, such as high efficiency, simple operation, shorter extraction time, and low cost [21,22]. The PILs were selected because of their advantageous property distinguishing them from other ILs. Firstly, the PILs have a strong ability to absorb microwave irradiation due to their highly polar nature [23,24]. Secondly, the synthesis of PILs is simple and easy, via equimolar mixing of acids and bases, possessing the advantages of low cost and easy preparation.

The aim of this work was, thus, to develop a fast and effective PIL MAE method for the extraction of rhein and emodin from *R. palmatum* L. As far as we know, PIL MAE methods used in the extraction of target products from *R. palmatum* L. are seldom reported. The effects of different extraction parameters, namely, the structure of the PILs, microwave irradiation power, PIL concentration, extraction time, and liquid–solid ratio, on the extraction properties were studied. The investigated technique showed the excellent extraction of rhein and emodin at a low cost.

2. Results and Discussion

2.1. The Choice of PIL

In this work, six PILs with different chemical structures were selected, and their extraction ability for rhein and emodin was studied. The results shown in Figure 1 indicate that the PILs with salicylate as the anion exhibited a lower extraction ability, e.g., $E([BzHim]MeSO_3) > E([BzHim]Sal)$. It is known that ILs with a high polarity have a stronger microwave absorption ability [25,26], leading to a higher extraction temperature and higher extraction efficiency. Therefore, the lower extraction ability of salicylate-based PILs may be attributed to their lower polarity. As reported in the literature [27–29], the ^{1}H-NMR chemical shifts (δ) of the hydrogen atom in the 2-position (H2) of imidazolium are related to the polarity of ILs, i.e., ILs with a stronger polarity have a higher chemical shift. As shown in Section 3.2, the δ values of H2 of ILs with $MeSO_3^-$ and p-TS$^-$ (p-toluenesulfonate) as anions were in the range of 9.136 to 9.296, and those of ILs with salicylate as the anion were within 8.541–8.629. This suggests that salicylate-based ILs have a lower polarity and, thus, exhibit poor extraction ability, which is in good agreement with the above observation (Figure 1). Furthermore, the extraction ability of [BHim]MeSO$_3$ was slightly higher than that of [BHim]p-TS, which may be ascribed to the fact that [BHim]MeSO$_3$ has a lower molecular weight and, thus, a higher molar concentration. Thus, [BHim]MeSO$_3$ was selected as the extraction solvent and used in the subsequent experiments.

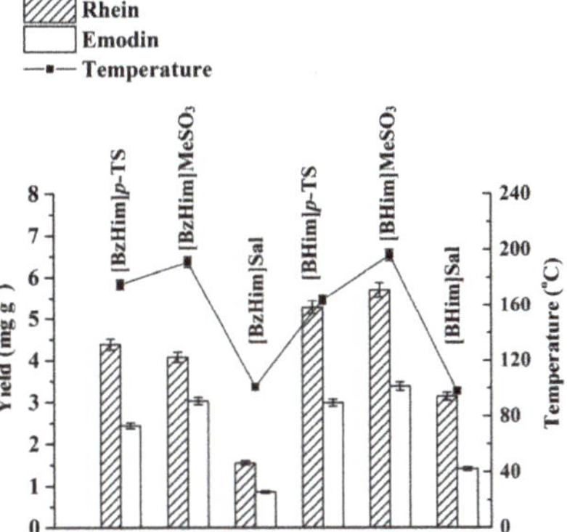

Figure 1. Extraction ability of different protic ionic liquids (PILs). Liquid–solid ratio, 20 g·g^{-1}; microwave irradiation power, 280 W; microwave irradiation time, 70 s; C_{IL} = 80%.

2.2. Selection of Extraction Conditions

In this work, the influence of the extraction conditions, such as liquid–solid ratio, extraction time, microwave irradiation power, and IL concentration on the extraction efficiency was studied systematically, and the results are shown in Figures 2 and 3. As shown in Figure 2A, the extraction efficiencies of rhein and emodin rose upon increasing the liquid–solid ratio from 10 to 40 g·g^{-1}, remained constant within 40–50 g·g^{-1}, and slightly decreased at 60 g·g^{-1}. Thus, a liquid–solid ratio of 40 g·g^{-1} was selected and used in the following studies.

The results shown in Figure 2B suggest that, within 30–50 s, a longer microwave irradiation time resulted in a higher extraction efficiency, and the extraction efficiency remained constant within 50–70 s. However, when the microwave irradiation time reached 80 s, the extraction efficiencies of rhein and emodin were rather poor, which can be attributed to the following fact: a longer microwave irradiation time (80 s) means that the target compounds must endure longer heating times, which may cause the partial decomposition of rhein and emodin. To confirm this deduction, the recoveries of rhein and emodin were determined (spiked level, 8.0 mg·g^{-1} for rhein and 4.0 mg·g^{-1} for emodin) under 80 s of microwave irradiation time. The experimental results indicated that the spiked recoveries of rhein and emodin were only 8.0% and 16.8%, respectively, which means that most of the two target compounds were decomposed. Based on the above discussion, 50 s of microwave irradiation time was chosen for the subsequent experiments.

The effect of microwave irradiation power on the extraction efficiency is illustrated in Figure 3A. As can be seen, the extraction efficiencies of rhein and emodin increased upon increasing the microwave irradiation power from 70 W to 280 W, which can be ascribed to the fact that high-power microwave irradiation leads to a higher extraction temperature and, thus, a higher extraction efficiency. When 350 W was adopted, the extraction efficiency decreased remarkably, which could also be attributed to the partial decomposition of rhein and emodin due to excessive temperature. Based on these results, 280 W was adopted as the microwave irradiation power for the following studies.

The experimental results illustrated in Figure 3B show that a higher IL concentration (60–85%) led to a higher extraction efficiency, and 90% IL resulted in a significant decrease in the extraction efficiency, which was ascribed to the excessive extraction temperature at 90% IL. In view of the fact that 80% IL exhibited a higher extraction ability, it was, thus, selected for the subsequent experiments.

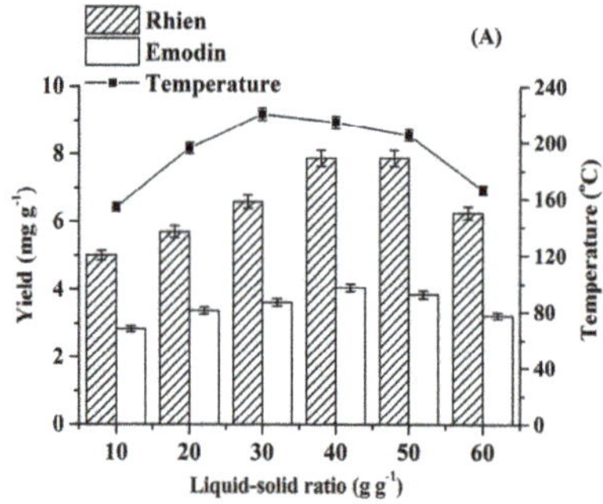
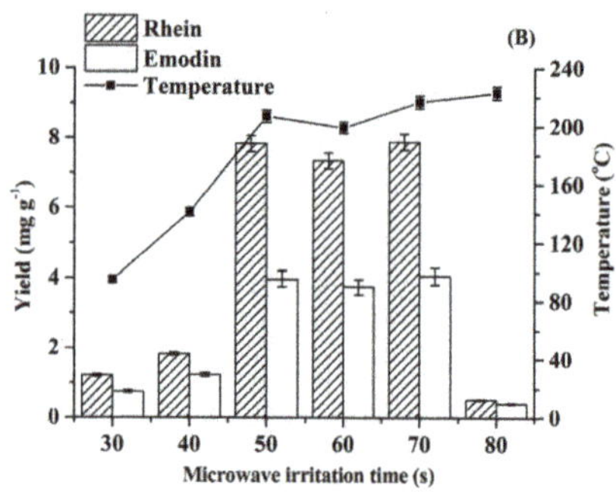

Figure 2. Influence of liquid–solid ratio (**A**) (microwave irradiation power, 280 W; microwave irradiation time, 70 s; C_{IL} = 80%) and microwave irradiation time (**B**) (microwave irradiation power, 280 W; liquid–solid ratio, 40 g·g^{-1}; C_{IL} = 80%) on the extraction efficiency.

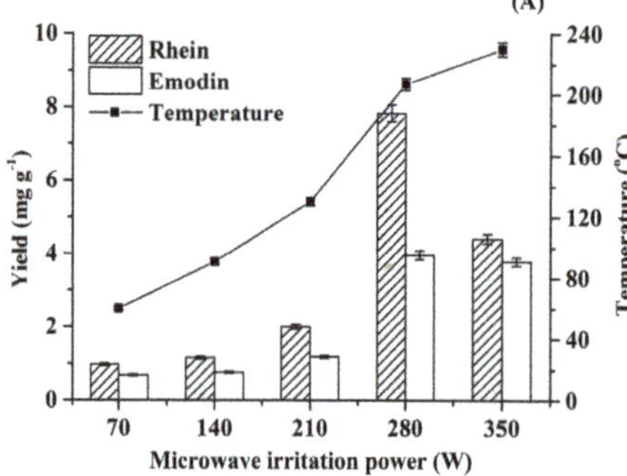
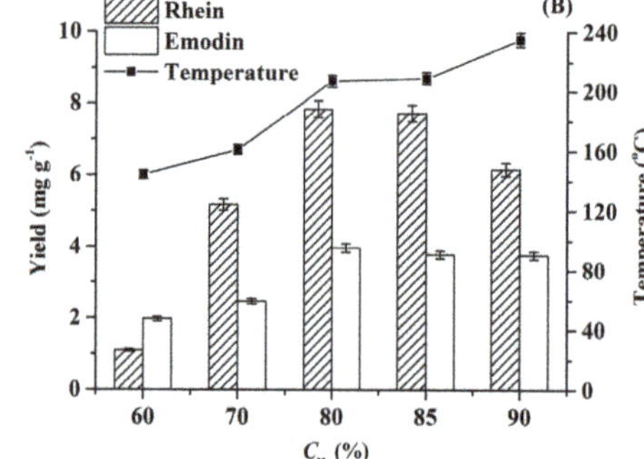

Figure 3. Influence of microwave irradiation power (**A**) (microwave irradiation time, 50 s; liquid–solid ratio, 40 g·g^{-1}; C_{IL} = 80%) and ionic liquid (IL) concentration (**B**) (microwave irradiation power, 280 W; liquid–solid ratio, 40 g·g^{-1}; microwave irradiation time, 50 s) on the extraction efficiency.

2.3. Comparison with Reported Methods

Recently, various extraction methods, such as UAE/LGH (where LGH is a solution of lactic acid, glucose, and water) [18], UAE/84% methanol [30] and HRE/methanol–trichloromethane [31] were used to extract rhein and emodin. Therefore, comparative experiments were conducted to investigate their ability for the extraction of rhein and emodin from the *R. palmatum* L. powder used in this work. Furthermore, the contrasting experiment of MAE/LGH was also done in this work. The results shown in Table 1 indicate that, compared with UAE and HRE, the MAE method is faster and more effective. It is worth mentioning that, under the selected experimental conditions, the extraction ability of MAE/[BHim]MeSO$_3$ was the best in terms of the contents of rhein (7.8 mg·g^{-1}) and emodin (4.0 mg·g^{-1}), better than the highest data obtained using HRE method (7.3 mg·g^{-1} and 3.5 mg·g^{-1}, respectively) [31]. Additionally, the product contents (rhein 2.2 mg·g^{-1} and emodin 0.87 mg·g^{-1}) of MAE/LGH were remarkably lower than those of MAE/[BHim]MeSO$_3$, which can be ascribed to the fact that [BHim]MeSO$_3$ is composed of ions and is more polar than LGH. Thus, the extraction temperature of MAE/[BHim]MeSO$_3$ was higher than MAE/LGH, resulting in the higher extraction ability of MAE/[BHim]MeSO$_3$. Above all, the extraction ability of PIL/MAE proposed by this work was faster and more effective than the previously reported extraction modes.

Table 1. Comparison between microwave-assisted extraction (MAE) with [BHim]MeSO$_3$ and other extraction methods. UAE—ultrasound-assisted extraction.

Extraction Method	Extraction Time	Content (mg·g^{-1})	
		Rhein	Emodin
UAE/LGH [a]	1.5 h	1.6	1.4
UAE/84% methanol [b]	33 min	1.9	1.1
HRE/methanol–trichloromethane [c]	2 h	7.3	3.5
MAE/LGH [d]	50 s	2.2	0.87
MAE/80% [BHim]MeSO$_3$ [e]	50 s	7.8	4.0

[a] LGH, lactic acid–glucose–water = 5:1:3. [a,b,c] Experiments were conducted under the conditions recommended by References [15,27,28], respectively. [d,e] Experiments were carried out under the conditions discussed above.

2.4. Recyclability of [BHim]MeSO$_3$ and the Purification of Rhein and Emodin

To reuse [BHim]MeSO$_3$, ethyl acetate is used to extract rhein and emodin from the [BHim]MeSO$_3$ solution after extraction. After filtration to remove *R. palmatum* L., the recovered [BHim]MeSO$_3$ can be used for the next extraction cycle without losing its extraction ability. Rhein and emodin in the ethyl acetate phase were purified using column chromatography using silica gel as the stationary phase and a mixture of ethyl acetate and petroleum ether as the eluent (30:70, *v/v*). The purity (determined by the HPLC method as described in Section 3) of the resultant rhein and emodin was 92.5% and 93.2%, respectively.

3. Materials and Methods

3.1. Reagents

Emodin (90%), rhein (98%), salicylic acid (99.5%), methanesulfonic acid (99%), and L-lactic acid (90%) were purchased from Aladdin Bio-Chem Technology Co., Ltd. (Shanghai, China). Furthermore, *p*-toluenesulfonic acid monohydrate (*p*-TSA·H$_2$O, 98%) was supplied by Tokyo Chemical Industry Co., Ltd. (Tokyo, Japan), while 1-benzylimidazole (98%) and 1-butylimidazole (98%) were obtained from Zhongkai Chem. Co., Ltd. (Changzhou, China). Ultrapure water (resistivity, 18.2 MΩ·cm) was used in the experiments (Aquapro purification system, Aquapro International Co., Ltd., Dover, DE, USA). All the other reagents used were of analytical grade unless stated otherwise. *R. palmatum* L. powder (100 mesh) was obtained from Xuanqing Biotechnology Co., Ltd. (Kunming, China).

3.2. Synthesis of PILs

To prepare 1-benzyl-3*H*-imidazolium salicylate ([BzHim]Sal), 0.1 mol of 1-benzylimidazole, 0.1 mol of salicylic acid, and 30 mL of methanol were mixed while stirring until a homogeneous solution was formed. After removing the solvent under vacuum, a light-yellow liquid was obtained. The PIL was identified by ^{1}H-NMR (400 MHz, DMSO-d_6 (hexadeuterodimethyl sulfoxide)) with the following chemical shifts (δ): 5.328 (s, 2H), 6.790–6.851 (m, 2H), 7.320–7.401 (m, 7H), 7.484 (s, 1H), 7.790–7.813 (m, 1H), 8.541 (s, 1H), and 12.725 (s, 1H).

The preparation of 1-benzyl-3*H*-imidazolium methanesulfonate ([BzHim]MeSO$_3$) followed the same procedure as for [BzHim]Sal described above, except methanesulfinic acid was used instead of salicylic acid. The product ([BzHim]MeSO$_3$) was a white solid with a melting point of 36–38 °C. The chemical shifts (δ) of ^{1}H-NMR (400 MHz, DMSO-d_6) were as follows: 2.349 (s, 3H), 5.451 (s, 2H), 7.380–7.430 (m, 5H), 7.707 (s, 1H), 7.806 (s, 1H), and 9.296 (s, 1H).

The preparation of 1-benzyl-3*H*-imidazolium *p*-toluenesulfonate ([BzHim]*p*-TS) followed the same procedure as for [BzHim]Sal described above, except *p*-toluenesulfonic acid was used instead of salicylic acid. The PIL, [BzHim]*p*-TS, was a white solid with a melting point of 106–108 °C. The chemical shifts

(δ) of ^{1}H-NMR (400 MHz, DMSO-d_6) were as follows: 2.287 (s, 3H), 5.438 (s, 2H), 7.106–7.125 (d, 2H), 7.395–7.422 (m, 5H), 7.479–7.499 (d, 2H), 7.697–7.705 (t, 1H), 7.794–7.801 (t, 1H), and 9.273 (s, 1H).

The synthesis of 1-butyl-3*H*-imidazolium-based PILs ([BHim]-based PILs) was similar to that of [BzHim]-based ones (equimolar mixing of 1-butylimidazole and a specific acid (*p*-toluenesulfonic acid, methanesulfinic acid, and salicylic acid)).

The PIL, [BHim]Sal, was a colorless liquid. The chemical shifts (δ) of ^{1}H-NMR (400 MHz, DMSO-d_6) were as follows: 0.852–0.888 (t, 3H), 1.171–1.245 (m, 2H), 1.692–1.765 (m, 2H), 4.082–4.118 (t, 3H), 6.756–6.815 (m, 2H), 7.299–7.342 (m, 1H), 7.420 (s, 1H), 7.553 (s, 1H), 7.787–7.810 (m, 1H), 8.629 (s, 1H), and 13.190 (s, 1H).

The PIL, [BHim]*p*-TS, was a white solid with a melting point of 58–59 °C. The chemical shifts (δ) of ^{1}H-NMR (400 MHz, DMSO-d_6) were as follows: 0.874–0.910 (t, 3H), 1.208–1.264 (m, 2H), 1.735–1.791 (m, 2H), 2.288 (s, 3H), 4.163–4.199 (t, 2H), 7.108–7.129 (d, 2H), 7.478–7.498 (d, 2H), 7.690 (s, 1H), 7.795 (s, 1H), and 9.136 (s, 1H).

The PIL, [BHim]MeSO$_3$, was a white solid with a melting point of 68–70 °C. The chemical shifts (δ) of ^{1}H-NMR (400 MHz, DMSO-d_6) were as follows: 0.880–0.916 (t, 3H), 1.218–1.274 (m, 2H), 1.746–1.802 (m, 2H), 2.339 (s, 3H), 4.175–4.212 (t, 2H), 7.698 (s, 1H), 7.804 (s, 1H), and 9.151 (s, 1H).

The chemical structures of the PILs, rhein, and emodin are shown in Figure 4.

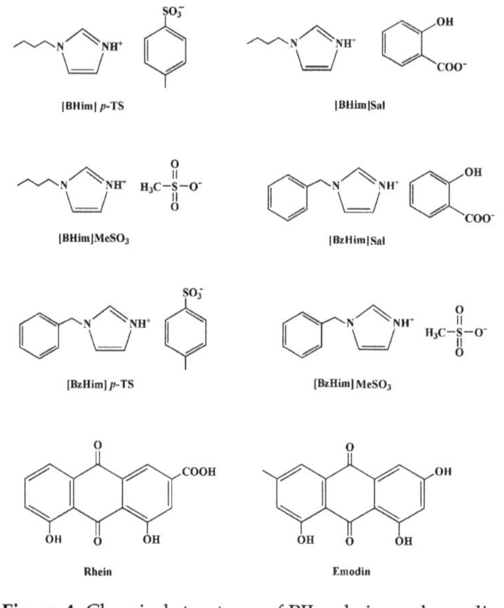

Figure 4. Chemical structures of PILs, rhein, and emodin.

3.3. Microwave-Assisted Extraction (MAE)

Briefly, a mixture of 0.1 g of *R. palmatum* L. powder and 2.0 g of PIL (80%, wt.%) was heated in a microwave oven (model G70F20CPIII-TK(W0), Guangdong Galanz Microwave Oven and Electrical Appliances Manufacturing Co., Ltd., Foshan, China). The microwave irradiation power and time were in the range of 70 W to 350 W and 30 s to 80 s, respectively. Extraction temperature (60 °C to 235 °C) was measured by a non-contact infrared thermometer (model, UT300A, Uni-Trend technology Co., Ltd., Dongguan, China). After extraction, the resultant mixture was diluted to 10 mL with ethanol and filtrated by a nylon membrane (pore size, 0.45 μm). The contents of emodin and rhein in ethanol solution were analyzed by an Agilent 1200 high-performance liquid chromatograph (HPLC, Agilent

Technologies, Santa Clara, CA, USA) equipped with a degasser, an autosampler, and a variable wavelength detector. The HPLC conditions were as follows: column temperature, 30 °C; injection volume, 5.0 µL; detection wavelength, 254 nm; separation column, ZORBAX Eclipse XDB-C18 column (4.6 mm × 150 mm, 5 µm, Agilent Technologies, Santa Clara, USA); flow rate, 1.0 mL·min^{-1}; mobile phase, a mixture of acetonitrile and 0.1% (*v/v*) acetic acid aqueous solution (40% (*v/v*) acetonitrile); run time, 15 min.

3.4. Heat Reflux Extraction (HRE)

This extraction procedure was conducted according to the standard method of the Chinese Pharmacopoeia [31]. Typically, 0.15 g of *R. palmatum* L. powder was mixed with 25 mL of methanol, and the resultant mixture was heated to reflux for 1.0 h. After cooling and filtration, 5.0 mL of the filtrate was withdrawn and evaporated under vacuum to remove the solvent; then, 10 mL of 8% HCl solution was added and subsequently sonicated for 2.0 min. After that, 10 mL of trichloromethane was added and heated to reflux for 1.0 h. After cooling, the trichloromethane phase was withdrawn, and the aqueous phase was washed three times, each time with 10 mL of trichloromethane. The trichloromethane phase was combined and evaporated under vacuum to remove the solvent. The resultant residue was dissolved with methanol and filtrated by a nylon membrane (pore size, 0.45 µm) before HPLC analysis.

3.5. Ultrasound-Assisted Extraction (UAE)

The UAE procedure was the same as that reported in the literature [30], with the following parameters: extraction solvent, 84% (*v/v*) methanol aqueous solution; liquid–solid ratio, 15:1 (mL/g); extraction time, 33 min; extraction temperature, 67 °C; extraction power, 250 W. After extraction, the filtrate was collected, and the residue was extracted again (two times) with the same volume of fresh solvent. The UAE method with LGH as the extraction solvent was conducted in a similar way under the conditions recommended by the reported work [18]. The contents of rhein and emodin were determined by the aforementioned HPLC method.

4. Conclusions

The present work developed a fast and effective MAE technique to extract rhein and emodin from *R. palmatum* L. using PILs as extraction solvents. The key advantage of PILs is that they are easy to prepare at low cost. Experimental results suggest that the PILs owning a higher polarity have a stronger microwave absorption ability and, thus, exhibit a higher extraction ability. The extraction ability of the PIL, [BHim]MeSO$_3$, was higher than that of DES (LGH), and the extraction ability of the proposed PIL-based MAE was superior to that of UAE and HRE. After extraction, the PIL, [BHim]MeSO$_3$, could be recycled via back extraction without losing its extraction ability. All the above results drive to the conclusion that PIL-based MAE is an efficient, rapid, simple, and economic method to extract rhein and emodin from *R. palmatum* L.

Author Contributions: Y.F., conceptualization and writing—original draft; Z.N., investigation; C.X., data curation; L.Y., writing—review and editing; T.Y., investigation.

Funding: This research was funded by the National Natural Science Foundation of China (21307028), the Foundation of Henan Province (182102310728), and the Program for Innovative Research Team of Henan Polytechnic University (T2018-3).

Conflicts of Interest: The authors declare no conflicts of interest.

References

1. Liu, Y.; Li, L.; Xiao, Y.Q.; Yao, J.Q.; Li, P.Y.; Yu, D.R.; Ma, Y.L. Global metabolite profiling and diagnostic ion filtering strategy by LC-QTOF MS for rapid identification of raw and processed pieces of *Rheum palmatum* L. *Food Chem.* **2016**, *192*, 531–540. [CrossRef] [PubMed]

2. Aichner, D.; Ganzera, M. Analysis of anthraquinones in rhubarb (*Rheum palmatum* and *Rheum officinale*) by supercritical fluid chromatography. *Talanta* **2015**, *144*, 1239–1244. [CrossRef] [PubMed]

3. Zhou, Y.; Zhang, J.; Tang, R.C.; Zhang, J. Simultaneous dyeing and functionalization of silk with three natural yellow dyes. *Ind. Crops. Prod.* **2015**, *64*, 224–232. [CrossRef]

4. Khan, S.A.; Shahid-ul-Islam; Shahid, M.; Khan, M.I.; Yusuf, M.; Rather, L.J.; Khan, M.A.; Mohammad, F. Mixed metal mordant dyeing of wool using root extract of *Rheum emodi* (Indian Rhubarb/Dolu). *J. Nat. Fibers* **2015**, *12*, 243–255. [CrossRef]

5. Huang, W.; Xue, A.; Niu, H.; Jia, Z.; Wang, J.W. Optimised ultrasonic-assisted extraction of flavonoids from Folium eucommiae and evaluation of antioxidant activity in multi-test systems in vitro. *Food Chem.* **2009**, *114*, 1147–1154. [CrossRef]

6. Sun, C.; Liu, H.Z. Application of non-ionic surfactant in the microwave-assisted extraction of alkaloids from *Rhizoma coptidis*. *Anal. Chim. Acta* **2008**, *612*, 160–164. [CrossRef]

7. Han, Y.; Yang, C.; Zhou, Y.; Han, D.; Yan, H. Ionic liquid–hybrid molecularly imprinted material–filter solid-Phase extraction coupled with HPLC for determination of 6-benzyladenine and 4-chlorophenoxyacetic acid in bean sprouts. *J. Agric. Food Chem.* **2017**, *65*, 1750–1757. [CrossRef]

8. Liang, R.; Bao, Z.; Su, B.; Xing, H.; Yang, Q.; Yang, Y.; Ren, Q. Feasibility of ionic liquids as extractants for selective separation of vitamin D3 and tachysterol3 by solvent extraction. *J. Agric. Food Chem.* **2013**, *61*, 3479–3487. [CrossRef]

9. Cao, Y.; Xing, H.; Yang, Q.; Bao, Z.; Su, B.; Yang, Y.; Ren, Q. Separation of soybean isoflavone aglycone homologues by ionic liquid-based extraction. *J. Agric. Food Chem.* **2012**, *60*, 3432–3440. [CrossRef]

10. Khezeli, T.; Daneshfar, A.; Sahraei, R. A green ultrasonic-assisted liquid-liquid microextraction based on deep eutectic solvent for the HPLC-UV determination of ferulic, caffeic and cinnamic acid from olive, almond, sesame and cinnamon oil. *Talanta* **2016**, *150*, 577–585. [CrossRef]

11. Amiri-Rigi, A.; Abbasi, S. Microemulsion-based lycopene extraction: Effect of surfactants, co-surfactants and pretreatments. *Food Chem.* **2016**, *197*, 1002–1007. [CrossRef] [PubMed]

12. Pawłowska, B.; Feder-Kubis, J.; Telesiński, A.; Biczak, R. Biochemical responses of wheat seedlings on the introduction of selected chiral ionic liquids to the soils. *J. Agric. Food Chem.* **2019**, *67*, 3086–3095. [CrossRef] [PubMed]

13. Tobiszewski, M.; Mechlinska, A.; Namiesnik, J. Green analytical chemistry—Theory and practice. *Chem. Soc. Rev.* **2010**, *39*, 2869–2878. [CrossRef] [PubMed]

14. Prat, D.; Pardigon, O.; Flemming, H.W.; Letestu, S.; Ducandas, V.; Isnard, P.; Guntrum, E.; Senac, T.; Ruisseau, S.; Cruciani, P.; et al. Sanofi's solvent selection guide: A step toward more sustainable processes. *Org. Process Res. Dev.* **2013**, *17*, 1517–1525. [CrossRef]

15. Procopio, A.; Alcaro, S.; Nardi, M.; Oliverio, M.; Ortuso, F.; Sacchetta, P.; Sindona, G. Synthesis, biological evaluation, and molecular modeling of oleuropein and its semisynthetic derivatives as cyclooxygenase inhibitors. *J. Agric. Food Chem.* **2009**, *57*, 11161–11167. [CrossRef] [PubMed]

16. Nardi, M.; Bonacci, S.; Cariati, L.; Costanzo, P.; Oliverio, M.; Sindona, G.; Procopio, A. Synthesis and antioxidant evaluation of lipophilic oleuropein aglycone derivatives. *Food Funct.* **2017**, *8*, 4684–4692. [CrossRef] [PubMed]

17. Nardi, M.; Bonacci, S.; De Luca, G.; Maiuolo, J.; Oliverio, M.; Sindona, G.; Procopio, A. Biomimetic synthesis and antioxidant evaluation of 3,4-DHPEA-EDA [2-(3,4-hydroxyphenyl) ethyl (3S, 4E)-4-formyl-3-(2-oxoethyl)hex-4-enoate]. *Food Chem.* **2014**, *162*, 89–93. [CrossRef] [PubMed]

18. Wu, Y.C.; Wu, P.; Li, Y.B.; Liu, T.C.; Zhang, L.; Zhou, Y.H. Natural deep eutectic solvents as new green solvents to extract anthraquinones from *Rheum palmatum* L. *RSC Adv.* **2018**, *8*, 15069–15077. [CrossRef]

19. Dai, Y.; van Spronsen, J.; Witkamp, G.J.; Verpoorte, R.; Choi, Y.H. Natural deep eutectic solvents as new potential media for green technology. *Anal. Chim. Acta* **2013**, *766*, 61–68. [CrossRef]

20. Wei, Z.F.; Wang, X.Q.; Peng, X.; Wang, W.; Zhao, C.J.; Zu, Y.G.; Fu, Y.J. Fast and green extraction and separation of main bioactive flavonoids from *Radix Scutellariae*. *Ind. Crops Prod.* **2015**, *63*, 175–181. [CrossRef]

21. Wang, S.Y.; Yang, L.; Zu, Y.G.; Zhao, C.J.; Sun, X.W.; Zhang, L.; Zhang, Z.H. Design and performance evaluation of ionic-liquids-based microwave-assisted environmentally friendly extraction technique for camptothecin and 10-hydroxycamptothecin from Samara of *Camptotheca acuminata*. *Ind. Eng. Chem. Res.* **2011**, *50*, 13620–13627. [CrossRef]

22. Chi, Y.; Zhang, Z.; Li, C.; Liu, Q.; Yan, P.; Welz-Biermann, U. Microwave-assisted extraction of lactones from *Ligusticum chuanxiong* Hort. using protic ionic liquids. *Green Chem.* **2011**, *13*, 666–670.

23. Greaves, T.L.; Drummond, C.J. Protic ionic liquids: Properties and applications. *Chem. Rev.* **2008**, *108*, 206–237. [CrossRef] [PubMed]

24. Hu, H.; Yang, H.; Huang, P.; Cui, D.; Peng, Y.; Zhang, J.; Lu, F.; Lian, J.; Shi, D. Unique role of ionic liquid in microwave-assisted synthesis of monodisperse magnetite nanoparticles. *Chem. Commun.* **2010**, *46*, 3866–3868. [CrossRef] [PubMed]

25. Arsalani, N.; Zare, P.; Namazi, H. Solvent free microwave assisted preparation of new telechelic polymers based on poly(ethylene glycol). *Express Polym. Lett.* **2009**, *3*, 429–436. [CrossRef]

26. Yang, F.; Gong, J.; Yang, E.; Guan, Y.; He, X.; Liu, S.; Zhang, X.; Deng, Y. Microwave-absorbing properties of room-temperature ionic liquids. *J. Phys. D Appl. Phys.* **2019**, *52*, 155302. [CrossRef]

27. Spange, S.; Lungwitz, R.; Schade, A. Correlation of molecular structure and polarity of ionic liquids. *J. Mol. Liq.* **2014**, *192*, 137–143. [CrossRef]

28. Chen, S.; Izgorodina, E.I. Prediction of ^{1}H NMR chemical shifts for clusters of imidazolium-based ionic liquids. *Phys. Chem. Chem. Phys.* **2017**, *19*, 17411–17425. [CrossRef]

29. Lin, S.T.; Ding, M.F.; Chang, C.W.; Lue, S.S. Nuclear magnetic resonance spectroscopic study on ionic liquids of 1-alkyl-3-methylimidazolium salts. *Tetrahedron* **2004**, *60*, 9441–9446. [CrossRef]

30. Zhao, L.C.; Liang, J.; Li, W.; Cheng, K.M.; Xia, X.; Deng, X.; Yang, G.L. The use of response surface methodology to optimize the ultrasound-assisted extraction of five anthraquinones from *Rheum palmatum* L. *Molecules* **2011**, *16*, 5928–5937. [CrossRef]

31. Chinese Pharmacopoeia Commission. *Pharmacopoeia of the People's Republic of China*, 10th ed.; Medical Science and Technology Press: Beijing, China, 2015; p. 23.

Sample Availability: Samples of the compounds are available from the authors.

Article

Direct Bioelectricity Generation from Sago Hampas by *Clostridium beijerinckii* SR1 Using Microbial Fuel Cell

Mohd Azwan Jenol [1], Mohamad Faizal Ibrahim [1] , Ezyana Kamal Bahrin [1], Seung Wook Kim [2,3] and Suraini Abd-Aziz [1,*]

[1] Department of Bioprocess Technology, Faculty of Biotechnology and Biomolecular Sciences, Universiti Putra Malaysia, 43400 UPM Serdang, Malaysia
[2] Department of Chemical and Biological Engineering, Korea University, Seoul 136-701, Korea
[3] Department of Chemistry, Faculty of Science and Technology, Universitas Airlangga, Surabaya 60115, Indonesia
* Correspondence: suraini@upm.edu.my; Tel.: +60-39769-1048

Received: 30 April 2019; Accepted: 24 May 2019; Published: 28 June 2019

Abstract: Microbial fuel cells offer a technology for simultaneous biomass degradation and biological electricity generation. Microbial fuel cells have the ability to utilize a wide range of biomass including carbohydrates, such as starch. Sago hampas is a starchy biomass that has 58% starch content. With this significant amount of starch content in the sago hampas, it has a high potential to be utilized as a carbon source for the bioelectricity generation using microbial fuel cells by *Clostridium beijerinckii* SR1. The maximum power density obtained from 20 g/L of sago hampas was 73.8 mW/cm^2 with stable cell voltage output of 211.7 mV. The total substrate consumed was 95.1% with the respect of 10.7% coulombic efficiency. The results obtained were almost comparable to the sago hampas hydrolysate with the maximum power density 56.5 mW/cm^2. These results demonstrate the feasibility of solid biomass to be utilized for the power generation in fuel cells as well as high substrate degradation efficiency. Thus, this approach provides a promising way to exploit sago hampas for bioenergy generation.

Keywords: sago hampas; starch; bioelectricity generation; *Clostridium beijerinckii*; microbial fuel cell

1. Introduction

Bio-electrochemical systems (BESs) are providing the innovative technologies that utilize the biological redox catalytic activity with the combination of abiotic electrochemical reactions [1,2], which are normally classified based on their applications, such as the generation of energy [3], chemicals and water treatment processes [4,5]. In fact, BESs used for energy generation are known as bio-electrochemical fuel cells, which can be further divided into enzymatic fuel cells (EFCs) and microbial fuel cells (MFCs). Microbial fuel cells (MFCs) represent a new platform of technology that converts the chemical energy in biomass into bioelectricity through metabolic activity of electrochemically active bacteria attached to the electrode [6]. Hence, the exploitation of bacteria as the electrocatalysts in MFCs has given the capability to directly generate electricity from a various type of substrate, including organic acids (acetate, butyrate, lactate), fermentable sugars (glucose, xylose) and carbohydrates (sucrose, starch) [7].

In spite of that, there is still more improvement required for reactor configurations and electrolyte design in MFCs, in order to enhance the effectiveness in terms of productivity as well as production cost [1]. According to Liu et al. [8], the substrate is indeed one of the most important biological factors affecting the overall performance of microbial fuel cells, including bioelectricity generation and operational cost. On the other hand, due to the versatility of fuel used in MFC, a novel approach has been introduced for MFCs as bioenergy production and biomass degradation technologies.

The utilization of biomass for energy generation has drawn considerable attention due to their abundance and ready availability [9]. In review, there are few sources of biomass that have been studied for the generation of bioelectricity, including potato waste [10], rice straw [11], wheat straw [12], and corn stover [13,14]. Du and Li [10] demonstrated bioelectricity generation in terms of current density from a mixture of cooked and uncooked potato, of which the maximum current density obtained was enhanced to 278 mA/m^2/d (10 fold increment). Hassan et al. [11] reported that the power density obtained from 1 g/L of initial rice straw concentration was 143 mW/m^2 with the coulombic efficiency (CE) in the range of 45.3–54.3%. In addition, a 123 mW/m^2 of power density was obtained by using a wheat straw hydrolysate, with the CE in the range of 15.5%–37.1% [12]. These findings provide evidence of the promising platform that utilizes the readily available biomass for bioelectricity generation. However, the use of other biomass should be investigated in order to provide a better understanding of and optimizing the generation of bioelectricity in MFCs.

In Malaysia, Sarawak is known as one of the largest sago starch exporters in the world, accounting for more than 25,000 mt/year of sago starch [15]. This figure is expected to increase by 15% to 20% every year [16]. It should be noted that the increase in a number of products will significantly increase the amount of waste produced from the sago processing mill, which subsequently is associated with environmental pollution problems if it is poorly handled. Sago hampas is a starchy lignocellulosic biomass produced during the extraction of sago starch which contains 58% starch, 23% cellulose, 9.2% hemicellulose and 3.9% lignin [17–19]. This sago hampas has a great potential to be utilized as a feedstock for bioenergy production. In MFCs, a starch component has been used for the generation of bioelectricity. Niessen et al. [20] has demonstrated the exploitation of *Clostridium beijerinckii* and *Clostridium butyricum* in bioelectricity generation using starch as an electron donor. The aforementioned authors reported that the current density obtained from *C. butyricum* and *C. beijerinckii* was 1.3 mA and 0.8–1.0 mA, respectively. On the other hand, the starch-based wastewater has been utilized as a fuel in MFCs, generating the highest voltage output of 281 with the power density of 139.8 mW/m^2 [21].

Therefore, this paper has aimed to utilize sago hampas as a substrate in the direct generation of bioelectricity by *C. beijerinckii* SR1 using a microbial fuel cell. Subsequently, the performance of sago hampas will be compared to the hydrolyzed sago hampas.

2. Results and Discussion

2.1. Characterization of Sago Hampas

In MFCs, the substrate is the most important fermentation factors affecting the production cost as well as bioelectricity generation [8]. Sago hampas is a great potential biomass produced after the extraction of starch and contains a significant amount of starch materials and fiber [17]. Table 1 illustrates the chemical composition of sago hampas used in this study for the generation of bioelectricity.

Table 1. Chemical composition of sago hampas.

Chemical Compositions (%)	This Study	Reference [17]	Reference [18]
Starch	58 ± 0.02	30–45	49.5
Cellulose	21	n.d	26
Hemicellulose	13.4	n.d	14.5
Lignin	5.4	n.d	7.5
Moisture	4.7 ± 0.42	5–7	n.d
Ash	3.13 ± 0.13	3–4	n.d
pH	4.49 ± 0.1	4.6–4.7	n.d

%—percentage in dry basis, n.d—not determined.

Sago hampas is a great potential biomass that consists both starch content as well as lignocellulosic materials. All the values (Table 1) except starch are almost comparable to the previous reports [17,18]. It is found that the amount of starch used in this study (58% on a dry basis) is slightly higher as compared to the previous study (30–50% on a dry basis). These differences in starch value are greatly dependent on the extraction protocol applied in the mills. Awg-Adeni et al. [17] have explained that higher amount of starch will be found in the sago hampas mainly due to the food grade starch demand, thus the factory has reduced the recycling process in the extraction to ensure the starch whiteness. It should be noted that this study was aimed to utilize the starch content in the biomass for the generation of bioelectricity.

2.2. Preliminary Experiment: Bioelectricity Generation from Commercial Starch

In this study, *Clostridium beijerinckii* SR1 was employed as a biocatalyst due to its high productivity and efficiency as a biocatalyst for hydrogen, which is the major electron donor in bioelectricity generation [20]. On the other hand, this bacterial strain is able to digest vast types of substrate, including low molecular compound like organic acids (acetate, butyrate), monosaccharides like glucose, and even starch [20,22,23]. It is also reported that *Clostridium* sp. have the ability to express α-amylase and glucoamylase, which are the important enzymes in degrading the starch. Figure 1 shows the growth profiling of *Clostridium beijerinckii* SR1.

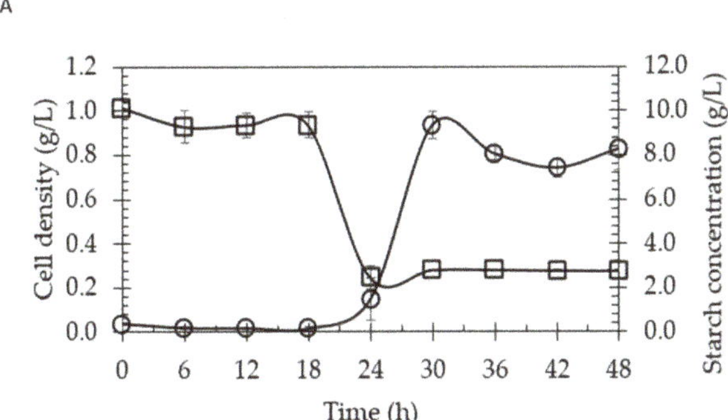

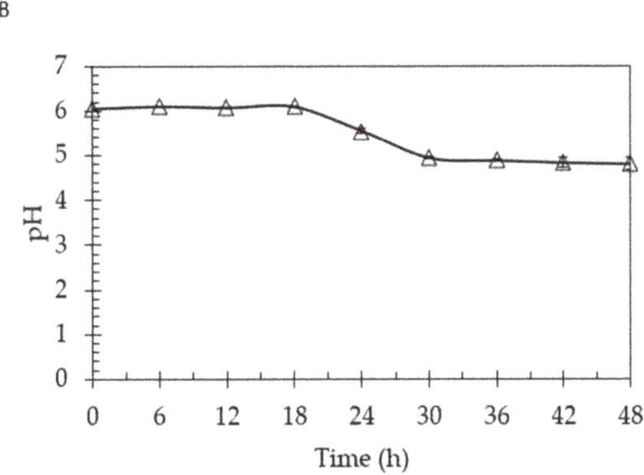

Figure 1. Profiling of *Clostridium beijerinckii* SR1 in 10 g/L of commercial starch. (**A**) The cell density (○) and commercial starch consumption (□) as a function of time. (**B**) pH changes.

Based on Figure 1, the highest cell density obtained was 0.93 ± 0.05 g/L at 30 h of fermentation. The pH dropped below pH 5 after 30 h, indicated by the acid production of this strain in the acidogenesis process. This situation caused the suppression of this bacteria to further grow when the accumulation of acids in the fermentation broth reached the threshold value just before the cells began to enter the stationary phase [24].

A preliminary experiment was conducted using commercial starch as an electron donor to demonstrate the capability of *C. beijerinckii* SR1 to generate the bioelectricity in MFCs system. Figure 2 illustrates the bioelectricity generation powered by the commercial starch as an electron donor by *C. beijerinckii* SR1 as a biocatalyst. Initially, when starch was introduced into the system, the circuit voltage (open circuit voltage (OVC)) of 26.4 ± 6.4 mV (0–6 h) was generated immediately and this might be due to the difference between anodic and cathodic potential of the two electrodes. Min et al. [25] explained that this situation might be due to the biological and chemical factors. Thereafter, the cell voltage (OVC) was increased and stabilized to 214.2 ± 7.3 mV (7–96 h) due to the biological activity of *C. beijerinckii* SR1. The voltage generation using two chambers of MFCs in the study was similar to the data showed by Min et al. [25].

A

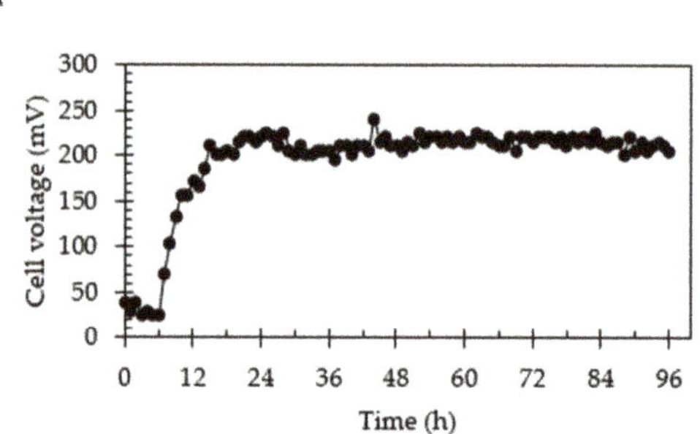

B

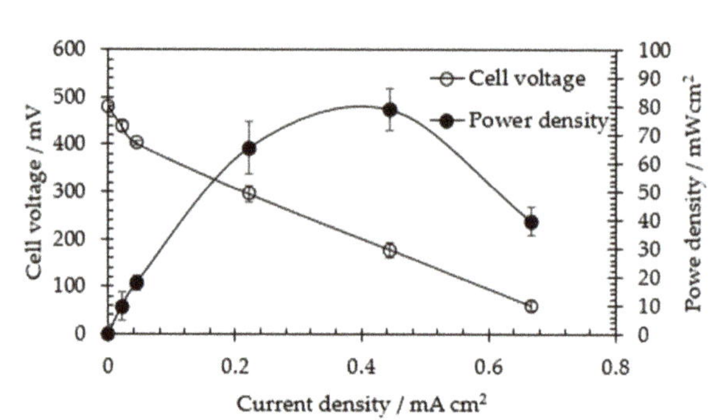

Figure 2. Bioelectricity generation from commercial starch by *Clostridium beijerinckii* SR1. (**A**) Voltage generation from 10 g/L of starch by *C. beijerinckii* SR1 as a function of time. (**B**) Power density and cell voltage generation as a function of current density.

The maximum power density obtained from the polarization curve (Figure 2B) was 78.92 ± 7.48 mW/cm^2 (at an applied current of 1.0 mA). The results showed that *C. beijerinckii* SR1 is a great potential biocatalyst capable of generating bioelectricity powered by starch. This situation is explained by the catalytic oxidation of hydrogen [20]. *Clostridium* sp. is a well-known biocatalyst that synthesizes biohydrogen via a ferredoxin-linked pathway of the butyric acid fermentation, then produces two moles of hydrogen per glucose unit in the fermentation of starch [26–28]. As explained by Niessen et al. [20], the main criteria of the selection of biocatalyst in the fuel cell are based on the efficiency of hydrogen synthesis as well as the ability to feed on the desired carbon source. In this study, we demonstrated the ability of *C. beijerinckii* SR1 in the generation of bioelectricity as well as the utilization of a macromolecular natural substrate (starch) and further investigated with solid biomass, which is sago hampas.

2.3. Direct Bioelectricity Generation by Clostridium beijerinckii SR1 Using Sago Hampas

Figure 3 exemplifies the bioelectricity generation as a function of cell voltage and power density by *C. beijerinckii* SR1 fueled by sago hampas. When the sago hampas was introduced, 33.8 ± 5.5 mV (0–11 h) of voltage output was generated out of the gate. Subsequently, the cell voltage was increased and stabilized at 211.7 ± 2.3 mV (12–96 h). The maximum power density recovered was 73.78 ± 2.6 mW/cm^2, at applied current 1.0 mA. The results obtained demonstrated that the performance of bioelectricity generation from sago hampas was almost comparable to commercial starch. This situation explained the potential of sago hampas as a replacement of commercial starch in bioelectricity generation using MFCs. The cell voltage output obtained in this study was slightly lower as compared to study made by Ahmed et al. [21]. The aforementioned authors used the starch-based wastewater as an electrolyte in a single chamber MFC with the maximum power density of 139.8 mW/m^2 and chemical oxygen demand (COD) removal of 86.4%. On the other hand, Neissen et al. [20] obtained 1.86 mW/cm^2 of power density by *C. butyricum* from 10 g/L of starch.

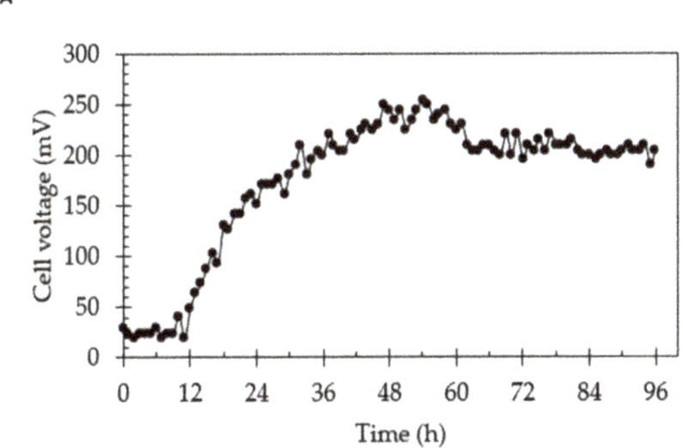
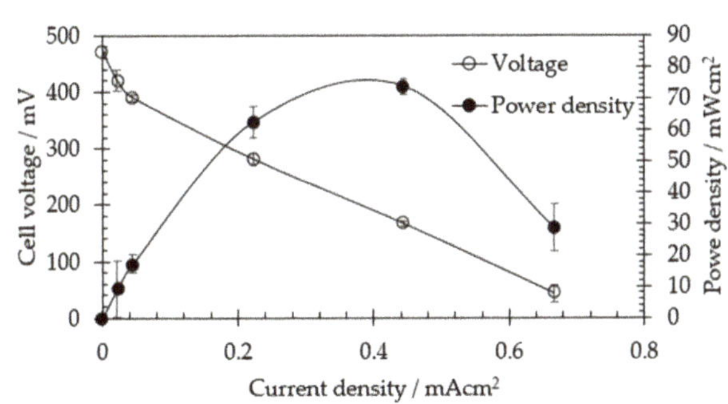

Figure 3. Bioelectricity generation from sago hampas by *Clostridium beijerinckii* SR1. (**A**) Voltage generation from 20 g/L of sago hampas by *C. beijerinckii* SR1 as a function of time. (**B**) Power density and cell voltage generation as a function of current density.

On the other hand, the result obtained from sago hampas was compared with the bioelectricity generation powered by hydrolyzed sago hampas, also known as sago hampas hydrolysate. Figure 4a illustrates the bioelectricity generation in the freshly inoculated anaerobic culture of *C. beijerinckii* SR1 containing sago hampas hydrolysate as a substrate. The concentration of 10 g/L of glucose content in the sago hampas hydrolysate was subjected as an initial carbon source. The maximum power density obtained from hydrolyzed sago hampas was 56.5 mW/cm^2, which was slightly lower than unhydrolyzed sago hampas. Based on the result obtained, it is suggested that there is no additional step (hydrolysis process) required in order to utilize the sago hampas as a substrate for the generation of bioelectricity.

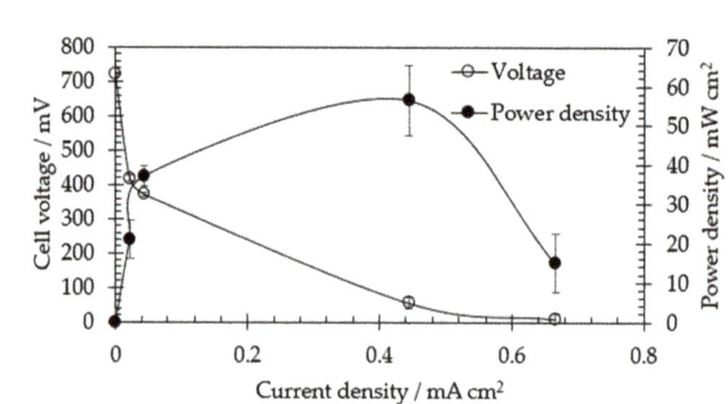

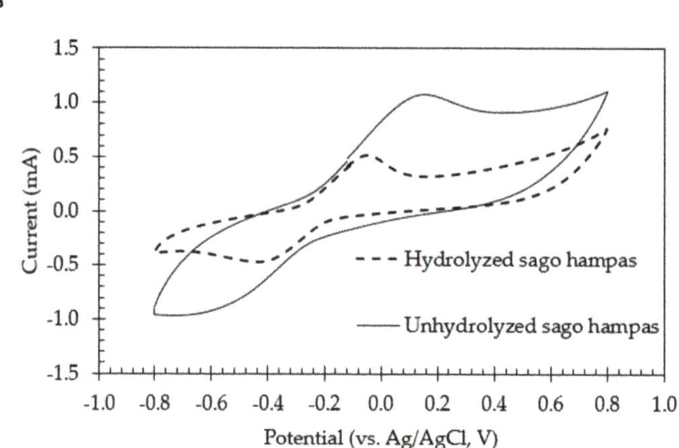

Figure 4. Generation from sago hampas hydrolysate by *Clostridium beijerinckii* SR1. (**A**) Power density and cell voltage generated from 10 g/L of glucose content in sago hampas hydrolysate as a function of current density. (**B**) Cyclic voltammetry profile of *Clostridium beijerinckii* SR1 recorded at 96 h after the start of anaerobic fermentation using unhydrolyzed and hydrolyzed sago hampas as substrate. The scan rate was 10 mV/s.

Based on the result obtained, the current obtained was 2.4 mA. The electrochemical behavior of MFC using sago hampas by *C. beijerinckii* SR1 was further evaluated using cyclic voltammetry (CV) that measures the potential differences across the interface as well as the redox of the component involved of the biochemical system. Figure 4b shows the redox potential signal obtained from hydrolyzed sago hampas and unhydrolyzed sago hampas. Based on the data obtained, unhydrolyzed sago hampas emitted a higher oxidation potential peak compared to hydrolyzed sago hampas, with the potential of 0.15 V and −0.02 V, respectively. The voltammogram shape obtained in this study was in agreement with Finch et al. [29]. The results obtained suggested that the electrochemical activity of *C. beijerinkcii*

SR1 might be due to the cell surface cytochrome(s) [29–31]. *C. beijerinckii* SR1 showed quasi-reversible redox reaction in CV with a sharper reduction peak as compared to the oxidation peak. It also appears that there was more efficient electron flow towards the bacteria cells.

Based on experimental results obtained, the performance of *C. beijerinckii* SR1 on the bioelectricity generation from commercial starch and sago hampas was subjected to the assessment of coulombic efficiency (CE). The CE was a function of either external circuit resistance or substrate concentration. The total starch consumed from sago hampas was 95.1% corresponding to a CE of 10.7%. This indicates that the major electron produced resulted from the degradation of starch and was not fully associated with power generation, but instead correlated to biomass production as well as fermentation products. *Clostridium* were known to be solvents (acetone–butanol–ethanol) and organic acids (acetic acid and butyric acid) producers. In this study, it is determined that this particular strain produced 4.02 ± 0.52 g/L of total solvents and 6.71 ± 0.45 g/L of total organic acids. This situation was in agreement with Lu et al. [32], which explained that the lower CE in bioelectricity generated by bacteria could be caused from the correlation of electron acceptor diffusion as well as other processes, including biomass production and fermentation. Logan et al. [33] have reported that the CE was diminished by several factors, including the competitive processes for the production of by-products as well as bacterial growth.

3. Materials and Methods

3.1. Substrate Collection and Preparation

The biomass used was sago hampas, which was collected from River Link Sago Resources Sendirian Berhad Company (Sdn. Bhd.), Mukah, Sarawak, Malaysia. Sago hampas was dried by using an oven dryer at 65 °C overnight. The dried sago hampas was then kept in the sealed plastic bags and stored at room temperature for further use.

3.2. Hydrolysis of Sago Hampas

Sago hampas containing 58% starch (dry basis) was subjected to the hydrolysis process [17,18]. A suspension of 7% (*w/v*) of sago hampas was boiled in 0.1 M KH_2PO_4 buffer solution (pH 4) for 15 min and subsequently cooled down to 60 °C at room temperature. A 5.56 U/mL of dextrozyme (Novozymes, Bagsværd, Denmark) with glucoamylase activity of 195.3 U/mL was added into the suspension. The mixture was incubated at 60 °C for 60 min in a water bath for the hydrolysis process. The mixture was continuously stirred to ensure the homogeneity of substrate end enzyme throughout the process. Then, the suspension was submerged into an ice-water bath to prevent further hydrolysis. The hydrolysate was then separated from the lignocellulosic fiber residual by using a filtration (100 mesh) and centrifugation (12,000 rpm for 15 min) technique. The hydrolysate obtained, referred as sago hampas hydrolysate was subjected to analytical procedures for its reducing sugars and glucose content. The hydrolysis yield (%) was calculated as:

$$\text{Glucose produced from sago hampas (g)/Dry sago hampas (g)} \times 100. \tag{1}$$

3.3. Bacterial Strain and Medium

The single culture, *Clostridium beijerinckii* SR1 employed was obtained from Madihah's laboratory culture collection, Universiti Teknologi Malaysia [34]. Aliquots of 1 mL of the stock culture was inoculated into 125 mL serum bottle, with 100 mL working volume of oxygen free and sterilized growth medium, pH 5.5. The medium [20] contained 10 g substrate (sago hampas), 5 g yeast extract, 10 g peptone, 0.5 g L-cysteine-HCl, 0.4 g $NaHCO_3$, 8 mg $CaCl_2$, 8 mg $MgSO_4$, 40 mg $KHSO_4$, 80 mg NaCl and 1 mg resazurin (per L). The pH was adjusted using 1.0 M of NaOH and HCl. The medium was purged with nitrogen gas for 15 min, and autoclaved at 121 °C for 15 min. It was incubated at

120 rpm and 37 °C for 24 h in a shaker incubator (Labwit, Victoria, Australia). The inoculum was freshly prepared for every bioelectricity generation in microbial fuel cells (MFCs).

3.4. MFC Construction and Operation

The MFCs used was the H-type, which consisted of conjoint two chambers; anode and cathode. An anaerobic chamber (anode) and aerobic chamber (cathode) were separated by the proton exchange membrane (PEM) (Nafion N117, Fuel Cell Earth, Massachusetts, United State of America). The electrodes (1.5 × 1.5 cm) used in the anode and cathode compartment were the carbon cloth (5% wetproofing) (Fuel Cell Earth, Massachusetts, USA) and 20% platinum on Vulcan—carbon cloth (Fuel Cell Store, USA), respectively. The pretreatment of electrodes and electrical connection were done based on that previously described by Oh et al. [35].

The MFCs were operated at a fixed external circuit resistance (300 Ω). Both chambers were filled with 200 mL working volume of fermentation medium. Sago hampas was added as an electron donor (commercial starch is used as a control) and 10% (*v/v*) of *C. beijerinckii* SR1 was inoculated into the anode compartment contained in the fermentation medium prior to the fermentation, simultaneously. Then, the nitrogen gas was purged into the anode compartment for 15 min to prepare the anaerobic condition. In the cathode compartment, 50 mM $K_3Fe(CN)_6$ was used as the terminal electron acceptor.

3.5. Calculation and Chemical Analysis

All the analysis samples were run for triplicates. Starch content was determined using the method explained by Nakamura [36]. The glucose concentration was determined by using high performance liquid chromatography (HPLC) (Jasco, Tokyo, Japan) equipped with Refractive Index (RI) detector and Rezex RCM-Monosaccharide (Reliability-Centered Maintenance) column (Phenomenex, USA) at 80 °C with deionized water as a mobile phase at flow rate of 0.6 mL/min. The solvent produced was analyzed using Gas chromatography (Shimazu, GC-17A, Kyoto, Japan) occupied with column BP-21 with helium as a carrier gas. Cell voltages (V) were measured across a fixed external circuit resistor (300 Ω) using a data acquisition system (LabJack U12 Series, LabJack Corporation, US) that recorded every 10 min. The polarization curve was determined by Autolab PGSTAT204 (Metrohm, Malaysia) imposing varying current (0–1.5 mA). The electron discharge pattern was also evaluated using Autolab PGSTAT204 (Metrohm, Malaysia) by studying the cyclic voltammetry (CV). CV was evaluated over a voltage range of −0.8 mV to + 0.8 mV at the scan rate of 10 mV/s. An Ag/AgCl electrode was used as a reference electrode, while an anode electrode and cathode electrode acted as a working and counter electrode, respectively. Power (P) density was calculated as:

$$P = I \times V \ (I = V/R) \tag{2}$$

where, I (A) is the current, V (V) is the voltage and R (Ω) is the external resistance. The power was normalized to the anode surface area [35,37]. The electrical charge (in coulombs) was calculated by integrating the current over time and the amount of substrate removal during a MFC operation [25,35,38]. Starch content was measured using iodine starch colorimetric explained by Nakamura [36].

The coulombic efficiency (CE) was calculated based on the total coulombs calculated by integrating the current over time (CP) and the theoretical amount of coulombs available from COD (CT) using Equation (3) as [32]:

$$CE = CP/CT \times 100\%. \tag{3}$$

The current over time (CP) is defined as $CP = \int I \, dt$, whereas the theoretical amount of coulombs (CT) is defined as $CT = FbV \, \Delta COD/M$, where F is the Faraday's constant (96485 C/mol of electrons), b is the number of moles of electrons produced per mol of substrates (b = 4), V (L) is the liquid volume, ΔCOD (g/L) is the concentration difference, and M is the molecular weight of substrate (M = 23).

4. Conclusions

This study evaluated the capability of the MFCs system in utilizing the solid biomass, sago hampas for the generation of bioelectricity by single culture, *Clostridium beijerinckii* SR1 in order to be one-step closer to better understanding of MFCs concept. In this study, three sources of substrate were tested, including commercial starch, unhydrolyzed and hydrolyzed sago hampas. The maximum power density generated by *C. beijerinckii* SR1 from unhydrolyzed sago hampas is 73.78 mW/cm^2 which is a comparable performance compared to commercial starch (78.92 mW/cm^2), while hydrolyzed sago hampas gave lower power density (56.5 mW/cm^2). These results indicate the value of unhydrolyzed sago hampas in being utilized directly as a carbon source without any further hydrolysis process for the generation of bioelectricity. It can be concluded that sago hampas has shown a great potential replacement for commercial starch in bioelectricity generation using MFCs. The improvement in terms of reactor configuration as well as electrodes manipulation could further enhance the generation of bioelectricity by using sago hampas as a feedstock in MFCs.

Author Contributions: Conceptualization, M.F.I., S.W.K. and S.A.-A.; Data Curation, M.A.J. and S.A.-A.; Formal Fnalysis, M.A.J.; Investigation, M.A.J.; Methodology, M.F.I.; Supervision, M.F.I., E.K.B., S.W.K. and S.A.-A.; Visualization, S.A.-A.; Writing—Original Draft, M.A.J.; Writing—Review and Editing, M.F.I., E.K.B. and S.A.-A. All authors have read and approved the final version.

Funding: This research was funded by the MyBrain15 from the Malaysian Higher Education Ministry.

Acknowledgments: The authors gratefully acknowledge all the members of the Environmental Biotechnology Research Group, Universiti Putra Malaysia for their help and support.

Conflicts of Interest: The authors declare no conflict of interest.

References

1. Ivars-Barceló, F.; Zuliani, A.; Fallah, M.; Mashkour, M.; Rahimnejad, M.; Luque, R. Novel Applications of Microbial Fuel Cells in Sensors and Biosensors. *Appl. Sci.* **2018**, *8*, 1184. [CrossRef]
2. Santoro, C.; Arbizzani, C.; Erable, B.; Ieropoulos, I. Microbial fuel cells: from fundamentals to applications. A review. *J. Power* **2017**, *356*, 225–244. [CrossRef] [PubMed]
3. Logan, B. Exoelectrogenic bacteria that power microbial fuel cells. *Nat. Rev. Microbiol.* **2009**, *7*, 375–381. [CrossRef] [PubMed]
4. Hampannavar, U.S.; Anupama, S.; Pradeep, N.V. Treatment of distillery wastewater using single chamber and double chambered MFC. *Int. J. Environ. Sci.* **2011**, *2*, 114–123.
5. Włodarczyk, P.P.; Włodarczyk, B. Microbial fuel cell with ni-co cathode powered with yeast wastewater. *Energies* **2018**, *11*, 3194. [CrossRef]
6. Logan, B.E. *Microbial Fuel Cell*; John Wiley & Sons: Hoboken, NJ, USA, 2008.
7. Pant, D.; Van Bogaert, G.; Diels, L.; Vanbroekhoven, K. A review of the substrates used in microbial fuel cells (MFCs) for sustainable energy production. *Bioresour. Technol.* **2010**, *101*, 1533–1543. [CrossRef]
8. Liu, Z.; Liu, J.; Zhang, S.; Su, Z. Study of operational performance and electrical response on mediator-less microbial fuel cells fed with carbon- and protein-rich substrates. *Biochem. Eng. J.* **2009**, *45*, 185–191. [CrossRef]
9. Pandey, P.; Shinde, V.N.; Deopurkar, R.L.; Kale, S.P.; Patil, S.A.; Pant, D. Recent advances in the use of different substrates in microbial fuel cells toward wastewater treatment and simultaneous energy recovery. *Appl. Energy* **2016**, *168*, 706–723. [CrossRef]
10. Du, H.; Li, F. Effects of varying the ratio of cooked to uncooked potato on the microbial fuel cell treatment of common potato waste. *Sci. Total Environ.* **2016**, *569–570*, 841–849. [CrossRef]
11. Hassan, S.H.A.; Gad El-Rab, S.M.F.; Rahimnejad, M.; Ghasemi, M.; Joo, J.H.; Sik-Ok, Y.; Kim, I.S.; Oh, S.E. Electricity generation from rice straw using a microbial fuel cell. *Int. J. Hydrogen Energy* **2014**, *39*, 9490–9496. [CrossRef]
12. Zhang, Y.; Min, B.; Huang, L.; Angelidaki, I. Generation of electricity and analysis of microbial communities in wheat straw biomass-powered microbial fuel cells. *Appl. Environ. Microbiol.* **2009**, *75*, 3389–3395. [CrossRef] [PubMed]

13. Zuo, Y.; Maness, P.C.; Logan, B.E. Electricity production from steam-exploded corn stover biomass. *Energy Fuels* **2006**, *20*, 1716–1721. [CrossRef]

14. Wang, X.; Feng, Y.; Wang, H.; Qu, Y.; Yu, Y.; Ren, N.; Li, N.; Wang, E.; Lee, H.; Logan, B.E. Bioaugmentation for electricity generation from corn stover biomass using microbial fuel cells. *Environ. Sci. Technol.* **2009**, *43*, 6088–6093. [CrossRef] [PubMed]

15. Awg-Adeni, D.S.; Abd-Aziz, S.; Bujang, K.B.; Hassan, M.A. Bioconversion of sago residue into value added products. *Afr. J. Biotechnol.* **2010**, *9*, 2016–2021.

16. Department of Agriculture Sarawak. Sarawak Agriculture Statistics. 2013. Available online: http://www.doa.sarawak.gov.my/modules/web/pages.php?mod=webpage&sub=page&id=712 (accessed on 16 August 2017).

17. Awg-Adeni, D.S.; Bujang, K.B.; Hassan, M.A.; Abd-Aziz, S. Recovery of glucose from residual starch of sago hampas for bioethanol production. *Biomed. Res. Int.* **2013**. [CrossRef] [PubMed]

18. Jenol, M.A.; Ibrahim, M.F.; Yee, P.L.; Md Salleh, M.; Abd-Aziz, S. Sago biomass as a sustainable source for biohydrogen production by Clostridium butyricum A1. *BioResources* **2014**, *9*, 1007–1026. [CrossRef]

19. Linggang, S.; Yee, P.L.; Wasoh, M.H.; Abd-Aziz, S. Sago pith residue as an alternative cheap substrate for fermentable sugars production. *Appl. Biochem. Biotechnol.* **2012**, *167*, 122–131. [CrossRef]

20. Niessen, J.; Schroder, U.; Scholz, F. Exploiting complex carbohydrates for microbial electricity generation—A bacterial fuel cell operating on starch. *Electrochem. Commun.* **2004**, *6*, 955–958. [CrossRef]

21. Ahmed, S.; Rozaik, E.; Abdelhalim, H. Performance of Single-Chamber Microbial Fuel Cells Using Different Carbohydrate-Rich Wastewaters and Different Inocula. *Pol. J. Environ. Stud.* **2016**, *25*, 503–510. [CrossRef]

22. Liu, J.; Guo, T.; Wang, D.; Ying, H. Clostridium beijerinckii mutant obtained atmospheric pressure glow discharge generates enhanced electricity in a microbial fuel cell. *Biotechnol. Lett.* **2015**, *37*, 95–100. [CrossRef]

23. Sun, G.; Thygesen, A.; Meyer, A.S. Acetate is a superior substrate for microbial fuel cell initiation preceding bioethanol effluent utilization. *Appl. Microbiol. Biotechnol.* **2015**, *99*, 4905–4915. [CrossRef] [PubMed]

24. Maddox, I.S.; Steiner, E.; Hirsch, S.; Wessner, S.; Gutierrez, N.A.; Gapes, J.R.; Schuster, K.C. The cause of "acid-crash" and "acidogenic fermentations" during the batch acetone-butanol-ethanol (ABE-) fermentation process. *J. Mol. Microbiol. Biotechnol.* **2000**, *2*, 95–100. [PubMed]

25. Min, B.; Kim, J.R.; Oh, S.E.; Regan, J.M.; Logan, B.E. Electricity generation from swine wastewater using microbial fuel cells. *Water Res.* **2005**, *39*, 4961–4968. [CrossRef] [PubMed]

26. Taguchi, F.; Chang, J.D.; Mizukami, N.; Saito-Taki, T.; Hasegawa, K.; Morimoto, M. Isolation of hydrogen-producing bacterium, *Clostridium beijerinckii* strain AM21B, from termites. *Can. J. Microbiol.* **1983**, *39*, 726–730. [CrossRef]

27. Taguchi, F.; Mizukami, N.; Saito-Taki, T.; Hasegawa, K. Hydrogen production from continuous fermentation of xylose during growth of *Clostridium* sp. strain No. 2. *Can. J. Microbiol.* **1995**, *41*, 536–540. [CrossRef]

28. Levin, D.B.; Pitt, L.; Love, M. Biohydrogen production: Prospects and limitations to practical application. *Int. J. Hydrogen Energy* **2004**, *29*, 173–185. [CrossRef]

29. Finch, A.S.; Mackie, T.D.; Sund, C.J.; Sumner, J.J. Metabolite analysis of *Clostridium acetobutylicum*: Fermentation in a microbial fuel cell. *Bioresour. Technol.* **2011**, *102*, 312–315. [CrossRef]

30. Myers, C.R.; Myers, J.M. Localization of Cytochromes to the Outer-Membrane of Anaerobically Grown Shewanella-Putrefaciens Mr-1. *J. Bacteriol.* **1992**, *174*, 3429–3438. [CrossRef]

31. Krige, A.; Sjöblom, M.; Ramser, K.; Christakopoulos, P.; Rova, U. On-line Raman spectroscopic study of cytochromes' redox state of biofilms in microbial fuel cells. *Molecules* **2019**, *24*, 646. [CrossRef]

32. Lu, N.; Zhou, S.; Zhuang, L.; Zhang, J.; Ni, J. Electricity generation from starch processing wastewater using microbial fuel cell technology. *Biochem. Eng. J.* **2009**, *43*, 246–251. [CrossRef]

33. Logan, B.E.; Hamelers, B.; Rozendal, R.; Schroder, U.; Keller, J.; Freguia, S.; Aelterman, P.; Verstraete, W.; Rabaey, K. Microbial fuel cells: Methodology and technology. *Environ. Sci. Technol.* **2006**, *40*, 5181–5192. [CrossRef] [PubMed]

34. Md Salleh, M. Application of Fed Batch System in The Production of Biobutanol by *Clostridium* sp. Using Lemongrass Leaves Hydrolysate. *Korean Biotechnol. Conf.* **2016**, *2016*, 12.

35. Oh, S.E.; Min, B.; Logan, B.E. Cathode performance as a factor in electricity generation in microbial fuel cells. *Environ. Sci. Technol.* **2004**, *38*, 4900–4904. [CrossRef] [PubMed]

36. Nakamura, L.K. *Lactorbacillus amylovorus*, a new starch-hydrolyzing species from cattle waste-corn fermentations. *Int. J. Syst. Bacteriol.* **1981**, *31*, 56–63. [CrossRef]

37. Oh, S.E.; Logan, B.E. Proton exchange membrane and electrode surface areas as factors that affect power generation in microbial fuel cells. *Appl. Microbiol. Biotechnol.* **2006**, *70*, 162–169. [CrossRef] [PubMed]
38. Liu, W.; Mu, W.; Liu, M.; Zhang, X.; Cai, H.; Deng, Y. Solar-induced direct biomass-to-electricity hybrid fuel cell using polyoxometalates as photocatalyst and charge carrier. *Nat. Commun.* **2014**, *5*, 1–8.

Sample Availability: Samples of the hydrolyzed and unhydrolyzed sago hampas available from the authors.

 *molecules*

Review

Improved Carotenoid Processing with Sustainable Solvents Utilizing Z-Isomerization-Induced Alteration in Physicochemical Properties: A Review and Future Directions

Masaki Honda [1,*], Hakuto Kageyama [1], Takashi Hibino [1], Yelin Zhang [2], Wahyu Diono [2], Hideki Kanda [2], Ryusei Yamaguchi [3], Ryota Takemura [4], Tetsuya Fukaya [4,5,*] and Motonobu Goto [2,*]

[1] Faculty of Science & Technology, Meijo University, Shiogamaguchi, Tempaku-ku, Nagoya 468-8502, Japan; kageyama@meijo-u.ac.jp (H.K.); hibino@meijo-u.ac.jp (T.H.)

[2] Department of Materials Process Engineering, Nagoya University, Furo-cho, Chikusa-ku, Nagoya 464-8603, Japan; zhang.yelin@c.mbox.nagoya-u.ac.jp (Y.Z.); wahyudiono@b.mbox.nagoya-u.ac.jp (W.D.); kanda.hideki@material.nagoya-u.ac.jp (H.K.)

[3] Technical Center, Nagoya University, Furo-cho, Chikusa-ku, Nagoya 464-8603, Japan; yamaguchi.ryusei@i.mbox.nagoya-u.ac.jp

[4] Innovation Division, Kagome Company, Limited, Nishitomiyama, Nasushiobara 329-2762, Japan; Ryota_Takemura@kagome.co.jp

[5] Institutes of Innovation for Future Society, Nagoya University, Furo-cho, Chikusa-ku, Nagoya 464-8603, Japan

* Correspondence: honda@meijo-u.ac.jp (M.H.); Tetsuya_Fukaya@kagome.co.jp (T.F.); goto.motonobu@material.nagoya-u.ac.jp (M.G.); Tel.: +81-52-838-2284 (M.H.); +81-287-36-2935 (T.F.); +81-52-789-3392 (M.G.)

Academic Editor: Monica Nardi
Received: 8 April 2019; Accepted: 5 June 2019; Published: 7 June 2019

Abstract: Carotenoids—natural fat-soluble pigments—have attracted considerable attention because of their potential to prevent of various diseases, such as cancer and arteriosclerosis, and their strong antioxidant capacity. They have many geometric isomers due to the presence of numerous conjugated double bonds in the molecule. However, in plants, most carotenoids are present in the all-*E*-configuration. (all-*E*)-Carotenoids are characterized by high crystallinity as well as low solubility in safe and sustainable solvents, such as ethanol and supercritical CO_2 (SC-CO_2). Thus, these properties result in the decreased efficiency of carotenoid processing, such as extraction and emulsification, using such sustainable solvents. On the other hand, Z-isomerization of carotenoids induces alteration in physicochemical properties, i.e., the solubility of carotenoids dramatically improves and they change from a "crystalline state" to an "oily (amorphous) state". For example, the solubility in ethanol of lycopene Z-isomers is more than 4000 times higher than the all-*E*-isomer. Recently, improvement of carotenoid processing efficiency utilizing these changes has attracted attention. Namely, it is possible to markedly improve carotenoid processing using safe and sustainable solvents, which had previously been difficult to put into practical use due to the low efficiency. The objective of this paper is to review the effect of Z-isomerization on the physicochemical properties of carotenoids and its application to carotenoid processing, such as extraction, micronization, and emulsification, using sustainable solvents. Moreover, aspects of Z-isomerization methods for carotenoids and functional difference, such as bioavailability and antioxidant capacity, between isomers are also included in this review.

Keywords: lycopene; β-carotene; astaxanthin; *E/Z*-isomerization; solubility; crystallinity; extraction; emulsification; micronization; supercritical CO_2

1. Introduction

Carotenoids are a class of lipid-soluble pigments responsible for the colors of plants, animals, and microorganisms [1–4]. Since the first structural elucidation of β-carotene by Kuhn and Karrer in the 1930s, approximately 1100 natural carotenoids have been reported so far [5]. Carotenoids can be classified into the following two groups based on their chemical composition: (1) carotenes, nonoxygenated molecules such as lycopene and β-carotene and (2) xanthophylls, molecules containing oxygen such as lutein and astaxanthin. (Figure 1) [4,6]. The daily consumption of carotenoid-rich foods, such as fruits and vegetables, is considered to be beneficial for human health because of their high antioxidant, anticancer, and antiatherosclerotic activities [7–9]. As carotenoids contain multiple conjugated double bonds, numerous geometric isomers are theoretically possible. While carotenoids in plants are accumulated predominantly as the all-*E*-configuration (Figure 1A–D), *Z*-isomers of carotenoids (Figure 1E,F) exist in abundance in the human body and in processed foods. For example, more than 50 and 30% of total lycopene are present as *Z*-isomers in human blood plasma and processed tomato products such as tomato sauce and tomato soup, respectively [10–12].

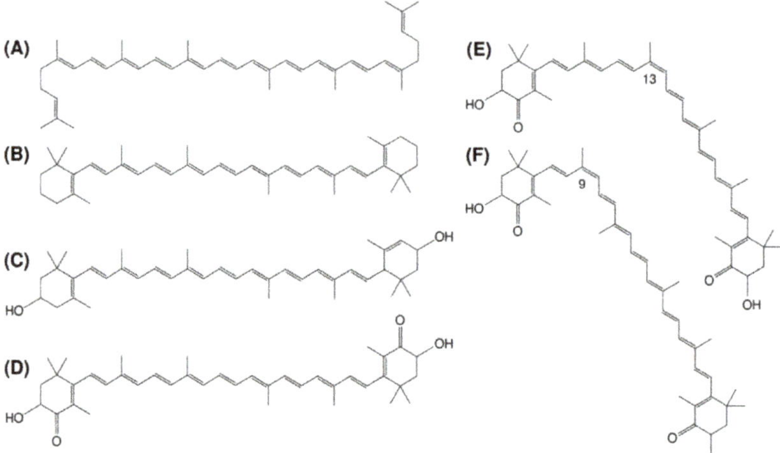

Figure 1. Chemical structures of (**A**) (all-*E*)-lycopene, (**B**) (all-*E*)-β-carotene, (**C**) (all-*E*)-lutein, (**D**) (all-*E*)-astaxanthin, (**E**) (13*Z*)-astaxanthin, and (**F**) (9*Z*)-astaxanthin.

Commercially available carotenoids are obtained by chemical syntheses or extraction from plants, photosynthetic bacteria, and microalgae. Generally, these carotenoids are in the all-*E*-configuration and the isomers are characterized by high crystallinity and low solubility in solvents [13,14]. Most carotenoid processing, such as extraction, micronization, and emulsification, employs a mediator—an organic solvent—to increase the processing efficiency. However, because of the physicochemical properties of (all-*E*)-carotenoids, processing efficiencies are low. Moreover, in recent years, there has been increased demand for the use of safe and sustainable solvents such as ethanol and supercritical CO_2 (SC-CO_2) for the processing of food components including carotenoids, i.e., environmentally benign processing using sustainable solvents is a topic of growing interest in both the research community and the food industry because of the growing awareness of the impact of solvents on energy usage, pollution, and their contribution to climate change and air quality [15–17]. However, since (all-*E*)-carotenoids have very low solubility in ethanol and SC-CO_2 [18–21], toxic organic solvents are used in many cases. Very recently, several studies demonstrated that *Z*-isomerization of carotenoids induces alteration in physicochemical properties, such as crystallinity and solubility. Namely, solubility in solvents including SC-CO_2 was dramatically improved and crystallinity was reduced by *Z*-isomerization. In addition, application of these alterations in carotenoid processing using the above safe and sustainable solvents has attracted attention. For example, Z-isomerization pretreatment significantly improved the extraction efficiency of lycopene contained in tomatoes and gac (*Momordica cochinchinensis* Spreng.) aril using organic solvents and SC-CO$_2$ [18,19].

In this review, the effect of Z-isomerization on the physicochemical properties of carotenoids and recent researches on carotenoid processing technology exploiting these characteristics are summarized and discussed. In addition, we also outline the typical methods for Z-isomerization of carotenoids and alterations in the bioavailability and functionality of carotenoids by Z-isomerization. Ample studies have demonstrated that Z-isomerization of carotenoids results in changes to bioavailability and antioxidant capacity, e.g., Z-isomers of lycopene and astaxanthin have greater bioavailability and show a higher antioxidant capacity than the all-*E*-isomers [22–25]. Thus, it is important to have a thorough understanding of the impact of *E/Z*-isomerization on functional changes of carotenoids.

2. Typical Methods for *Z*-Isomerization of Carotenoids

As the method for Z-isomerization of carotenoids, heat treatment, microwave treatment, light irradiation, electrolysis treatment, and catalytic treatment have been well documented to date (Table 1). (all-*E*)-Carotenoids, e.g., lycopene and astaxanthin, were efficiently isomerized to the Z-isomers by heating in organic solvents, especially alkyl halides such as dichloromethane (CH$_2$Cl$_2$) and dibromomethane (CH$_2$Br$_2$) [26–28]. The all-*E*-isomers were also thermally Z-isomerized in the presence of vegetable oils, animal fats, and SC-CO$_2$ [12,21]. These results indicate that for Z-isomerization of (all-*E*)-carotenoids, it is important that the carotenoid is in a dissolved state. Microwave heating also promoted Z-isomerization [29–32], and several studies indicated the increased efficiency compared to conventional heating [29,30]. In the microwave treatment of (all-*E*)-lycopene-rich tomato oleoresin, the total Z-isomer content reached 65.9 ± 2.7% with 4-min irradiation at 2450 MHz frequency and 500 W power, and the temperature of the oleoresin reached 136.7 ± 6.6 °C at that time, while the total Z-isomer content with conventional oil bath heating at 140 °C for 5 min was 50.8 ± 3.2% [29]. Light irradiation also caused *E/Z*-isomerization of carotenoids. When carotenoids were directly exposed to light, Z-isomers of carotenoids isomerized to the all-*E*-isomers [33,34]. On the other hand, when light irradiation was carried out with photosensitizers, such as chlorophyll *a* and erythrosine, Z-isomerization of (all-*E*)-carotenoids was promoted [35,36]. For example, when purified (all-*E*)-lycopene dissolved in hexane in the presence of erythrosine was irradiated at 480–600 nm for 1 h, the proportion of lycopene Z-isomers reached over 80% [36]. A few studies have demonstrated that electrolysis treatment promoted Z-isomerization of (all-*E*)-carotenoids such as β-carotene and canthaxanthin [37,38]. This electrochemical method shows very high efficiency and can prevent thermal degradation of carotenoids, e.g., approximately 50% of (all-*E*)-canthaxanthin was converted to the Z-isomers during 4–6 min of bulk electrolysis at 4 °C [37]. Catalytic Z-isomerization of (all-*E*)-carotenoids have been traditionally conducted using iodine [39–41]. More recently, it was reported that disulfide compounds [29,42,43], isothiocyanates, carbon disulfide [42], iron(III) chloride [44], titanium tetrachloride [45], and iodine-doped titanium dioxide [46] induced Z-isomerization of carotenoids. For example, when iron(III) chloride, which is usually used as a food additive for iron fortification, was used, greater isomerization (79.9%) could be attained by a 3-h reaction at 60 °C, with almost no degradation of lycopene [44]. Although catalyst utilization for carotenoid isomerization is a very efficient method, most catalysts, such as iodine and heavy metals, exert a negative impact on the human body and the environment. Hence, in industrial view, the discovery and use of low toxicity and environmentally sustainable catalysts, e.g., plant-derived natural catalysts such as disulfide compounds and isothiocyanates, will be of great importance [29,42,43,47–49]. As shown in Table 1, each Z-isomerization method has several advantages and disadvantages; consequently, it is necessary to select the appropriate Z-isomerization method according to the circumstances.

Table 1. Summary of representative methods for Z-isomerization of carotenoids and their advantages and disadvantages.

Method	Reported Carotenoid	Advantage	Disadvantage	Reference
Heat treatment	Lycopene, β-carotene, astaxanthin, lutein, etc.	• Simple and conventional method • Requires minimal amount of additive	• Can cause thermal degradation	[12,21,26–28]
Microwave treatment	Lycopene, β-carotene, astaxanthin, lutein, etc.	• Simple and fast (few minutes) method • Requires minimal amount of additive	• Can cause thermal degradation • Difficult to control final product quality • High cost of instrumentation	[29–32]
Light irradiation	Lycopene, β-carotene, lutein, etc.	• Rapid method • Non-thermal process • Low energy usage	• Can cause light degradation • Need to add photosensitizers • Need to remove photosensitizers if toxic ones used • High cost of some photosensitizers	[26,33–36]
Electrolysis treatment	β-Carotene, 8′-apo-β-caroten-8′-al, canthaxanthin	• Simple and highly efficient method • Non-thermal process • Can prevent degradation during processing	• High cost of instrumentation • Need to remove electrolyte substances if toxic ones used	[37,38]
Catalytic treatment	Lycopene, β-carotene, astaxanthin, zeaxanthin, etc.	• Simple and highly efficient method • Can prevent degradation during the processing • Low energy usage	• Need to remove catalysts if toxic ones used • Can promote degradation in some catalysts • High cost of some catalysts	[29,39–46]

3. Effect of Z-Isomerization on Bioavailability and Functionality of Carotenoids

Ample studies have addressed the alterations in bioavailability and functionality, such as antioxidant, anticancer, and antiatherosclerotic activities, of carotenoids by Z-isomerization (Table 2). Further, the alterations differed among carotenoids. Z-Isomers of lycopene and astaxanthin showed greater bioavailability than the all-E-isomers [22,23,25,50–53]. For example, investigation of the effect of red tomato juice (90% all-E-isomer lycopene) and *tangerine* tomato juice (94% Z-isomer lycopene) ingestion on plasma lycopene concentrations revealed that lycopene from *tangerine* tomato juice showed approximately 8.5-fold greater bioavailability than lycopene from red tomato juice [22]. On the other hand, some carotenoid Z-isomers, such as β-carotene and lutein, may be less bioavailable than the all-E-isomers [54–62]. In general, the bioavailability of carotenoids is very low because they are strongly bound to the food matrix and because of their physicochemical characteristics, such as low solubility, high crystallinity, and lipophilicity [4,13,14]. Thus, to improve the bioavailability of carotenoids contained in fruits and vegetables, physical treatments, such as high-pressure homogenization and ultrasound treatment, have been traditionally studied [63]. In some carotenoids, such as lycopene and astaxanthin, by combining chemical approaches such as Z-isomerization treatment and the above physical approaches, it may be possible to further improve the bioavailability.

Depending on the assay method, many studies have reported that Z-isomers of carotenoids have equal or higher antioxidant capacity compared with the all-E-isomers [24,25,46,63–70]. Böhm et al. (2002) [64] reported that Z-isomers of lycopene exhibited approximately 1.2 times higher antioxidant capacities than the all-E-isomer in the Trolox equivalent antioxidant capacity (TEAC) assay. In heme-induced peroxidation of linoleic acid in mildly acidic emulsions, which mimics postprandial lipid oxidation in the gastric compartment (MbFe^{III}-LP) assay, (5Z)-lycopene showed approximately 3 times higher antioxidant capacity than the all-E-isomer [24]. In contrast, when antioxidant capacity was evaluated by the TEAC assay, (9Z)-zeaxanthin exhibited lower capacity than the all-E-isomer [64]. The degree of antioxidant capacity varied among Z-isomers, e.g., that of lutein isomers was higher in the order of 13Z-isomer > 9Z-isomer > all-E-isomer with the ferric reducing antioxidant power (FRAP) assay [62]. Carotenoid Z-isomers are likely to be superior to the all-E-isomers in preventative effects on atherosclerosis, cancer, and inflammatory [71–76]. For example, (9Z)-β-carotene contained in the microalgae *Dunaliella* sp. showed higher antiatherogenic [71] and antiatherosclerotic [72,73] activities than the all-E-isomer in mouse experiments. (9Z)-Canthaxanthin isolated from *Dietzia* sp. exhibited higher proapoptotic activity than the all-E-isomer in THP-1 macrophages [74]. Nakazawa et al. (2009) [75] reported that Z-isomers of fucoxanthin had greater anticancer activity than the all-E-isomer in human promyelocytic leukemia (HL-60) and colon cancer (Caco-2) cells. Very recently, Yang et al. (2019) [76] showed that Z-isomers of astaxanthin, especially the 9Z-isomer, exhibited greater antiinflammatory effect than the all-E-isomer by downregulating proinflammatory cytokines COX-2 and TNF-α gene expression, which was evaluated in a Caco-2 cell monolayer model. As another notable functional change by carotenoid Z-isomerization, Z-isomers of β-carotene, which is a very important retinol precursor with a high conversion rate, showed lower conversion efficiencies to retinol than the all-E-isomer [77,78]. The antiaging activity would also differ among astaxanthin isomers. Namely, the median lifespan of *Caenorhabditis elegans* fed with 9-Z-, 13-Z-, and all-E-isomers was observed to increase by 59.39%, 24.99%, and 30.43%, respectively [79]. Moreover, Fenni et al. (2019) [80] reported that lycopene isomers exert similar biological functions in adipocytes, linked to their ability to transactivate PPARγ. Since Z-isomerization had "positive" or "negative" effects on the bioavailability and functionality of carotenoids (Table 2), it is important to have a detailed understanding of the impact of E/Z-isomerization on corresponding functional changes.

Table 2. Comparison of the bioavailability and functionality of all-*E*- and *Z*-isomers of carotenoids.

Carotenoid	Bioavailability	Antioxidant Capacity	Other Functionality
Lycopene	• $E^{\,a} < Z^{\,b}$ (Oral study in humans) [22,50] • $E < Z$ (Oral study in ferrets) [51] • $E < Z$ (Caco-2 cell model) [52] • $E < Z$ (Diffusion model) [53]	• $E \leq Z$ (TEAC assay) [24,64] • $E < Z$ (LPSC assay) [24] • $E \leq Z$ (MbFe^{III}-LP assay) [24] • $E \approx Z$ (FRAP assay) [24]	Antiobesity activity: • $E \approx 5Z$ (Adipocyte model) [80]
α-Carotene	–	• $13'Z > E \approx 9'Z > 9Z \approx 13Z$ (TEAC assay) [64]	–
β-Carotene	• $E > 9Z$ (Oral study in humans) [54–58] • $E > Z$ (Oral study in ferrets) [59] • $E > Z$ (Oral study in gerbils) [60] • $E > Z$ (Caco-2 cell model) [61] • $E < Z$ (Digestion model) [81]	• $E < Z$ (Oral study in rats) [65] • $E \approx Z$ (TEAC assay) [64] • $E \approx 9Z \approx 13Z > 15Z$ (TEAC assay) [66] • $E \approx Z$ (FRAP assay) [66] • $E \approx 9Z \approx 13Z > 15Z$ (CL assay) [66]	Antiatherogenesis activity: • $E < 9Z$ (Oral study in mice) [71] Antiatherosclerosis activity: • $E < 9Z$ (Oral study in mice) [72,73]
Astaxanthin	• $E \leq Z$ (Oral study in humans) [23] • $E < Z$ (Caco-2 cell model) [25] • $E < Z$ (Digestion model) [25]	• $E < Z$ (DPPH assay) [46,67] • $E < Z$ (ORAC assay) [46] • $E < Z$ (PLC assay) [46] • $E < Z$ (Enzyme activity assay) [25] • $E < Z$ (Lipid- peroxidation assay) [67]	Antiinflammatory activity: • $E < Z$ (Caco-2 cell model) [76] Antiaging activity: • $9Z > E > 13Z$ (*Caenorhabditis elegans* model) [79]
Canthaxanthin	–	• $E < 9Z$ (DPPH assay) [68] • $E < 9Z$ (Fluorescence assay) [68]	Proapoptotic activity: • $E < 9Z$ (THP-1 macrophage model) [74]
Fucoxanthin	–	• $E < Z$ (DPPH assay) [69] • $13Z$ and $13'Z > E > 9'Z$ (DPPH assay) [70] • $13Z$ and $13'Z > E > 9'Z$ (Superoxide-detection assay) [70] • $9'Z > E > 13Z$ and $13'Z$ (ABTS assay) [70] • $9'Z > E > 13Z$ and $13'Z$ (Hydroxyl radical-scavenging assay) [70]	Anticancer activity: • $E < Z$ (Caco-2 cell model) [75] • $E < Z$ (HL-60 cell model) [75]
Lutein	• $E > Z$ (Caco-2 cell model) [62] • $E < Z$ (Digestion model) [62]	• $E < Z$ (FRAP assay) [62] • $13'Z > E \approx 9Z$ (DPPH assay) [62] • $13'Z > E \approx 9Z$ (ORAC assay) [62] • $E \approx Z$ (CAA assay) [62]	–
Zeaxanthin	–	• $E \approx 13Z > 9Z$ (TEAC assay) [64]	–

^{a} all-*E*-isomer of carotenoid. ^{b} *Z*-isomer of carotenoid.

4. Effect of *Z*-Isomerization on Physicochemical Properties of Carotenoids

The *Z*-isomerization of (all-*E*)-carotenoids induces change in physicochemical properties such as color, solubility, crystallinity, melting point, and stability. *Z*-Isomerization of carotenoids resulted in a shift in absorption to a shorter wavelength and a reduction in the molar extinction coefficient and color value [27,46,82,83]. For example, Jing et al. (2012) [83] reported that maximum absorption wavelengths of (all-*E*)-, (9*Z*)-, and (13*Z*)-β-carotene were 451.4, 446.4, and 439.1 nm, respectively. The molar extinction coefficients of (all-*E*)-, (9*Z*)-, and (13*Z*)-lycopene at the maximum absorption wavelengths were 182×10^3, 164×10^3, and 137×10^3 M^{-1} cm^{-1}, respectively [27]. In fact, tomatoes rich in (all-*E*)-lycopene show a red color, whereas tomatoes rich in the *Z*-isomers, known as *tangerine* tomatoes, show an orange color [22].

Several studies reported that *Z*-isomers of carotenoids had much higher solubility than the all-*E*-isomers in organic solvents, oils, and SC-CO$_2$ [13,14,18–21,84,85]. Although the solubility of (all-*E*)-lycopene in ethanol, acetone, ethyl acetate, and hexane was 0.6, 42.7, 145.3, and 25.6 mg/mL, respectively, that of lycopene containing 75.6% *Z*-isomers was 2401.7, 3702.9, 3961.1, and 3765.2 mg/mL, respectively [13]. Namely, in the case of ethanol, which is frequently used for food processing such as extraction and purification, the solubility of lycopene *Z*-isomers was over 4000 times higher than that of the all-*E*-isomer. Also, in SC-CO$_2$, the solubility of (9*Z*)-β-carotene was nearly four times higher than that of the all-*E*-isomer [84], and lycopene *Z*-isomers also showed higher solubility than the all-*E*-isomer [19,21]. The increased solubility of carotenoids by *Z*-isomerization is likely to be associated with changes in bioavailability. Generally, carotenoids are absorbed from the duodenum and prior to the absorption they are incorporated into bile acid micelles [86]. Thus, since carotenoid *Z*-isomers may have higher solubility in bile acid than all-*E*-isomers, they are preferentially incorporated into enterocytes and show higher bioavailability [51,87]. On the other hand, *Z*-isomers of β-carotene exhibit lower bioavailability in humans than the all-*E*-isomer [54–58]. Several proteins, which are temporarily present at the apical membrane of the duodenum, mediate selective carotenoid uptake [86]. Therefore, β-carotene *Z*-isomers may be efficiently incorporated into bile acid micelles due to their high solubility, but may have lower transport efficiency in the duodenum than the all-*E*-isomer. In vitro experiments using Caco-2 cells strongly support the above hypothesis. Namely, *Z*-isomers of lycopene and astaxanthin showed higher cellular uptake efficiency than the all-*E*-isomers [25,52], while the opposite result was obtained for β-carotene [61]. Similarly, Yang et al. (2018) [62] reported that in vitro experiments using a digestion model shown higher bioaccessibility of lutein *Z*-isomers than the all-*E*-isomer, while a Caco-2 cell monolayer model revealed lower bioavailability.

Z-Isomerization of carotenoids affects the crystallinity. Murakami et al. (2017) [13] and Honda et al. (2018) [14] experimentally revealed that increases in the *Z*-isomer content of lycopene, β-carotene, and astaxanthin was related to a reduction in crystallinity, i.e., scanning electron microscopy (SEM), differential scanning calorimetry (DSC), and powder X-ray diffraction (XRD) analyses clearly demonstrated that (all-*E*)-carotenoids were present in a crystal state, while *Z*-isomers were present in an amorphous state. Carotenoids have multiple conjugated double bonds in the molecule, resulting in strong π–π stacking interactions between molecules. For this reason, carotenoids have high crystallinity. However, the presence of *Z*-isomers is suggested to lead to enormous steric hindrance and decrease the potential attractive π–π forces, thus affecting the crystallinity [13,88]. Generally, carotenoids in fresh plants occur predominantly in the (all-*E*)-configuration, and (all-*E*)-carotenoids are present in the crystal state. On the other hand, some plants, such as *tangerine* tomato and peach palm (*Bactris gasipaes* Kunth), contain high amounts of carotenoid *Z*-isomers that are present in an oily aggregate form [22,89]. Similarly, 9*Z*-isomer-rich β-carotene contained in *Dunaliella* was in the oily form [90].

The melting point of carotenoids was altered by *Z*-isomerization, i.e., increases in the *Z*-isomer content were associated with a lower melting point [13,14,85,91]. For example, the melting point of (all-*E*)-lycopene and lycopene containing 23.8, 46.9, and 75.6% *Z*-isomers was 174.4, 173.7, 170.0, and 162.3 °C, respectively, as measured by DSC [13].

The stability of carotenoids varies among isomers, i.e., (all-*E*)-carotenoids had higher stability than the *Z*-isomers. Several studies investigated the stability of carotenoid isomers using a Gaussian program and revealed that Gibbs free energy differed among the isomers [82,92–94]. For example, Takehara et al. (2015) [93] reported that the relative stability of lycopene isomers was in the following order; all-*E*-isomer ≈ 5*Z*-isomer > 9*Z*-isomer > 13*Z*-isomer > 15*Z*-isomer, and Guo et al. (2008) [94] reported that the relative stability of β-carotene isomers was in the following order; all-*E*-isomer > 9*Z*-isomer > 13*Z*-isomer > 15*Z*-isomer > 7*Z*-isomer ≈ 11*Z*-isomer. Murakami et al. (2018) [33] experimentally confirmed the above for lycopene. Furthermore, they investigated the stability of lycopene isomers against light irradiation, and the stability was in the following order; all-*E*-isomer ≈ 5*Z*-isomer > 9*Z*-isomer > 13*Z*-isomer > multi-*Z*-isomers. As for lycopene *Z*-isomers, the 5*Z*-isomer showed the highest stability against heat and light. In addition, (5*Z*)-lycopene would have higher antioxidant capacity [24] and bioavailability [95] compared with the all-*E*-isomer and possibly the 9*Z*- and 13*Z*-isomers. Therefore, it is important to develop a facile procedure for lycopene isomerization from the all-*E*-isomer to the 5*Z*-isomer.

The differences in physicochemical properties between (all-*E*)-carotenoids and *Z*-isomers are summarized in Table 3. A systematic understanding of these carotenoid properties is likely to be important in the analysis, processing, and so on.

Table 3. Differences in physicochemical properties between (all-*E*)-carotenoids and *Z*-isomers.

Color Value	Solubility	Crystallinity	Melting Point	Stability
E [a] > *Z* [b]	*E* < *Z*	*E* > *Z*	*E* > *Z*	*E* > *Z*

[a] all-*E*-isomer of carotenoid. [b] *Z*-isomer of carotenoid.

5. Improvement of Carotenoid Processing Efficiency by Z-Isomerization

In recent years, due to the discovery of altered physicochemical properties of carotenoids by *Z*-isomerization, efforts to improve the efficiency of carotenoid processing by exploiting these alterations has attracted attention. In particular, carotenoid processing using a safe and sustainable solvent—SC-SO$_2$—as a mediator is being actively studied. Since natural carotenoids, the all-*E*-isomer, exhibit very low solubility in SC-SO$_2$, there is a high hurdle for its industrial use in carotenoid processing. However, utilizing alterations in the physical properties by *Z*-isomerization represents a breakthrough. In this section, we introduce recent studies of carotenoid processing (extraction, micronization, and emulsification) utilizing alterations in solubility and crystallinity of carotenoids by *Z*-isomerization.

5.1. Improvement of Carotenoid Extraction

Generally, commercially available natural carotenoids, which are obtained from plants and microorganisms by solvent extraction and utilized for supplements, food colorants, and cosmetics, are very expensive [96–99]. This is because carotenoids in plants and microorganisms accumulate predominantly in the all-*E*-configuration, whose isomers have low solubility in solvents, resulting in very low extraction efficiencies. For example, extraction of lycopene from tomato pulp with ethanol and SC-CO$_2$ showed a recovery of only 6.3 and 6.5%, respectively [19]. However, when the extractions were conducted after *Z*-isomerization treatment, the recovery was notably improved to 75.9 and 27.6%, respectively [19]. More specifically, the total *Z*-isomer content of lycopene in tomato pulp was 8.8%, whereas it increased to 75.7% by heating at 150 °C for 1 h with a small amount (1 wt%) of olive oil. After ethanol extraction of lycopene from the *Z*-isomer-rich tomato pulp, the obtained extract had a very high *Z*-isomer content (93.5%), while almost all lycopene in the extraction residue was the all-*E*-isomer. These results strongly indicated that lycopene *Z*-isomers have higher solubility in solvents than the all-*E*-isomer; thus, the extraction efficiency was improved. In addition, since the *Z*-isomer content of carotenoids in the obtained extract was improved by *Z*-isomerization pretreatment, the treatment is effective not only for the production of carotenoid concentrates but also for increasing the bioavailability and functionality of carotenoids (Figure 2). The improved extraction efficiency was also confirmed in gac (*M. cochinchinensis* Spreng.) aril [18]. Gac is a tropical vine originating from South and South-East Asia and belongs to the Cucurbitaceae family, and the aril (seed membrane) contains a very high amount of lycopene [100,101]. Since gac aril contains a large amount of oil (18–34% of dry weight) rich in lycopene, lycopene is often obtained by press extraction with the oil [102]. Although more than 90% of lycopene exists as the all-*E*-isomer in gac aril, the total *Z*-isomer content increased by 58.5% with microwave irradiation at 1050 W for 60 s. When lycopene was obtained by press extraction with gac oil from non-microwave pretreated and treated gac aril, lycopene contents in the obtained oils were 160.6 and 1365.9 mg/100 g, respectively. Thus, *Z*-isomers of carotenoids show higher solubility in oils than the all-*E*-isomer. Moreover, *Z*-isomerization pretreatment of gac aril was also effective for lycopene extraction using ethanol and SC-CO_2. For example, when lycopene was extracted using SC-CO_2 from the non-treated gac aril, the lycopene content in the extract was only 76.6 mg/100 g, whereas *Z*-isomerization pretreatment by microwave irradiation resulted in a lycopene content of 342.0 mg/100 g. As the extraction efficiency of carotenoids is improved by *Z*-isomerization pretreatment, the development of efficient *Z*-isomerization methods for carotenoids in plants is very important in the future. On the other hand, several plants and microalgae such as *tangerine* tomato and *Dunaliella* contain a high amount of carotenoid *Z*-isomers [22,71,72]. Thus, carotenoids should be efficiently extracted using these raw materials. In fact, Gamlieli-Bonshtein et al. (2002) [84] reported that (9*Z*)-β-carotene in *Dunaliella* exhibited nearly 4 times higher extraction efficiency by SC-CO_2 than the all-*E*-isomer. Pretreatments of samples by physical and chemical approaches such as grinding, osmotic shock, bead-beating, high-pressure homogenization, and enzymatic treatment are effective in releasing carotenoids from complex matrices, and have been performed in basic and applied studies [103,104]. On the other hand, *Z*-isomerization pretreatment is a new technology reported very recently. By combing traditional physical and chemical pretreatments and *Z*-isomerization pretreatment, further improvement of carotenoid extraction can be expected. In addition, when the *Z*-isomerization pretreatment is used in combination with several extraction technique, such as pulsed electric field-assisted extraction, microwave-assisted extraction, and ultrasonic-assisted extraction, synergistic effects are expected [105–109].

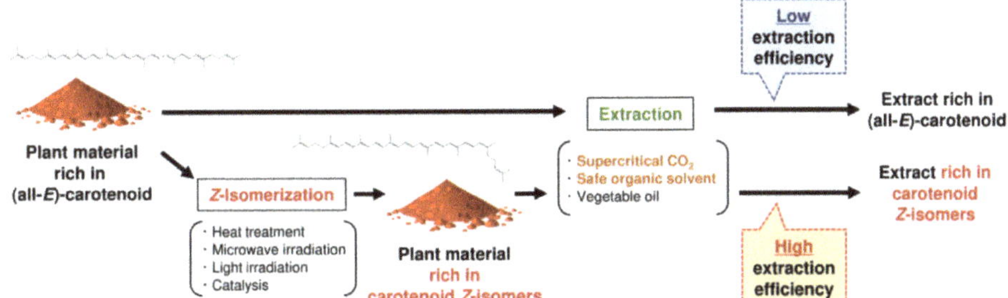

Figure 2. Schematic chart showing extraction of lycopene from plant material by solvents in the case of using (all-*E*)-lycopene and lycopene *Z*-isomers as the raw materials [18,19].

5.2. Improvement of Carotenoid Micronization

Ample studies have reported that micronization of carotenoids results in their increased bioavailability [110,111]. Generally, carotenoid micronization is conducted by milling, grinding, and chemical precipitation [112–114]. However, there are some concerns regarding the above conventional methods, as carotenoids are easily decomposed by friction heat and oxygen contact. In addition, when using chemical processes, toxic organic solvents may remain. Thus, in recent years, micronization of carotenoids using SC-CO_2 has attracted increasing attention. Since CO_2 is nontoxic and has a low critical temperature (T_c = 31.1 °C), it is suitable for heat-sensitive materials such as carotenoids, and SC-CO_2 is easily separated from the products along with the toxic organic solvent [115,116]. To the best of our knowledge, improved micronization efficiency of carotenoids utilizing alterations in the physicochemical properties by Z-isomerization has been reported only for the method using SC-CO_2 [117]. Particle micronization techniques using SC-CO_2, supercritical antisolvent (SAS), solution-enhanced dispersion by supercritical fluids (SEDS), rapid expansion of supercritical solutions (RESS), gas antisolvent (GAS), supercritical fluid extraction of emulsions (SFEE), and particles from gas saturated solutions (PGSS) have been well-documented [118–122]. Several studies have examined the micronization of carotenoids using the above techniques; however, there was difficulty in obtaining nano-sized carotenoid particles [123–125]. For example, Tavares-Cardoso et al. (2009) [125] conducted micronization of (all-*E*)-β-carotene using a SAS process under various conditions; however, nano-sized β-carotene particles could not be obtained. This is likely because of the high crystallinity of carotenoids. On the other hand, Kodama et al. (2018) [117] successfully prepared nano-sized lycopene by SEDS precipitation using lycopene Z-isomers as the raw material. Namely, when using (all-*E*)-lycopene as the raw material, particles having an average size of 3.6 μm were obtained, whereas when using lycopene containing 97.8% Z-isomers, uniformly sized particles of an average size of 75 nm were obtained (Figure 3). The reason why nanoparticles were successfully formed from Z-isomers is due to the low crystallinity compared with the all-*E*-isomer. In addition, little has been reported on carotenoid micronization using RESS precipitation: the substance, which must be reduced in size, is dissolved in pure SC-CO_2 and then the solution is suddenly depressurized through a nozzle and expands inside a chamber under lower pressure. This would be because carotenoids have extremely low solubility in pure SC-CO_2. However, as Z-isomers of carotenoids have relatively high solubility in SC-CO_2 [18,19,21,84], the Z-isomers would successfully form nano-sized particles by RESS precipitation, representing a micronization method without the use of organic solvents.

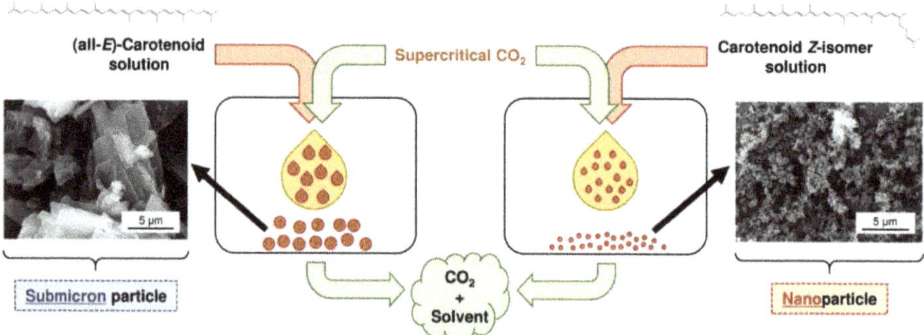

Figure 3. Schematic chart showing preparation of lycopene particles with supercritical CO_2 (solution-enhanced dispersion by supercritical fluids), using (all-*E*)-lycopene and lycopene Z-isomers as the raw materials [117].

5.3. Improvement of Carotenoid Emulsification

In recent years, as carotenoids are safe natural pigments that have health enhancing effects, their demand by the food industry is continuously increasing [126,127]. However, the low water solubility of carotenoids has made their use problematic for food formulations, limiting the favorable effects of carotenoids. Furthermore, the low water solubility of carotenoids reduces their bioavailability [128,129]. Therefore, improved dispersibility in water by emulsification is very important for the food industry and acts to increase their bioavailability. It is preferred that the suspended preparation contains nano-sized particles for higher dispersibility and bioavailability [111,130]. To obtain nanosuspensions of carotenoids, the following emulsification–evaporation technique is frequently used [131–133]: (1) Dissolution of carotenoids in an organic phase; (2) Distribution processing of the solution with water containing an emulsifier; (3) Solvent evaporation under reduced pressure. In this technique, it is important to select an appropriate distribution processing method, e.g., ultrasound treatment, high-speed homogenization, high-pressure homogenization, and microfluidizer treatment [131–136]. In addition, the selection of a solvent that can dissolve the target carotenoid is also a very important factor to efficiently produce carotenoid emulsions. However, since the degree of carotenoid solubility in safe and sustainable solvents, such as ethanol and supercritical SC-CO_2, is very low [14,18,19,21,84], toxic solvents are used in many cases. To improve the emulsification efficiency of carotenoids using the sustainable solvent SC-CO_2, Ono et al. (2018) [20] focused on increased carotenoid solubility in solvents by Z-isomerization. Namely, they investigated the impact of Z-isomer content on the production of β-carotene suspensions by the emulsification–evaporation technique. As the organic phase, they used SC-CO_2 (Figure 4). When β-carotene rich in Z-isomers (79.1% of total β-carotene) was used as the raw material, the encapsulated β-carotene content was notably increased compared with the all-*E*-isomer. For example, the encapsulated β-carotene content was 21.2 times higher after emulsification treatment by ultrasound at 45 kHz for 60 min. In addition, when (all-*E*)-β-carotene was used as the raw material, the mean particle size of the obtained suspension was approximately 700 nm, whereas that of β-carotene rich in Z-isomers was approximately 100 nm. Thus, Z-isomerization treatment before distributed processing is effective for the preparation of carotenoid suspensions by the emulsification–evaporation technique. However, the storage stability of a Z-isomer-rich β-carotene suspension was lower than that of all-*E*-isomer-rich one, possibly due to increases in the contact area with oxygen as the particle size decreased [20]. For practical application of this suspension preparation technique, establishment of a method to increase the storage stability of carotenoid Z-isomers is essential.

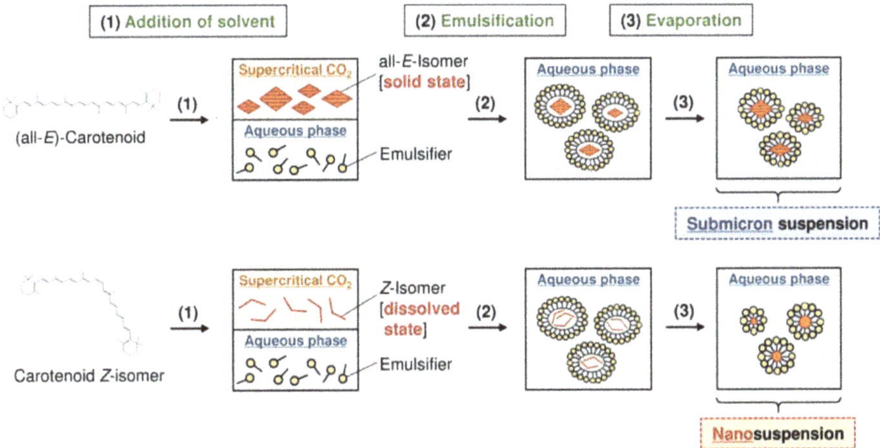

Figure 4. Schematic chart showing preparation of β-carotene suspensions by emulsification–evaporation technique with SC-CO_2, using (all-*E*)-β-carotene and β-carotene Z-isomers as the raw materials [20].

6. Conclusions and Future Perspectives

This review summarizes alterations in the physicochemical properties (color value, solubility, crystallinity, melting point, and stability) of carotenoids by Z-isomerization and their application for carotenoid processing (extraction, micronization, and emulsification), specifically using a green and sustainable solvent—SC-CO_2—and presents typical Z-isomerization methods and the effect of Z-isomerization on the bioavailability and functionality of carotenoids. As the method for Z-isomerization of carotenoids, heat treatment, microwave treatment, light irradiation, electrolysis treatment, and catalytic treatment have been well reported. Since these Z-isomerization methods have several advantages and disadvantages, it is necessary to select the appropriate Z-isomerization method according to the circumstances. Ample studies have demonstrated that Z-isomerization of carotenoid affected the bioavailability, antioxidant capacity, and functionalities such as anticancer activity and antiinflammatory activity and often offered positive impacts on human. The Z-isomerization also induces changes in the physicochemical properties of carotenoids, such as solubility and crystallinity. Namely, the solubility in organic solvents, SC-CO$_2$, and oils dramatically is enhanced and crystallinity is reduced by Z-isomerization. Since the (all-*E*)-carotenoid, which is a predominant isomer in plants and synthetic ingredients, has very low solubility in SC-CO$_2$, its industrial use in carotenoid processing faces a very high hurdle. However, it is highly expected that this impediment could be improved by utilizing the alterations in physicochemical properties of carotenoids by Z-isomerization. Carotenoid processing utilizing Z-isomerization and the expected application of Z-isomer-rich carotenoid materials are summarized in Figure 5. Plants and microalgae rich in carotenoid Z-isomers would be applicable as raw materials for the efficient extraction of carotenoids using solvents such as SC-CO$_2$, for use in health foods, food colorants, and animal feed. The obtained extract rich in carotenoid Z-isomers is expected to be applied to the production of supplements and food colorants with high carotenoid bioavailability and functionality. When safe and sustainable extraction solvents, such as ethanol and supercritical CO$_2$, are employed, the value of the extract is anticipated to increase. Furthermore, utilization of carotenoid Z-isomer-rich extracts as the raw material is expected to increase the production and quality of nano-sized carotenoids and carotenoid emulsions. The obtained nano-sized carotenoids and carotenoid emulsions rich in Z-isomers are expected to be utilized as supplements, food colorants, and cosmetics. In addition, alterations in the physicochemical properties of carotenoids by Z-isomerization may be beneficial for the production of microcapsules prepared using carotenoid-containing liposomes. The studies on increasing efficiency of carotenoid processing by Z-isomerization pretreatment has just started in recent years. Thus, there is still considerable room for the development of this research field. Fundamental study of this technology will be actively conducted in the future and practical applications are expected.

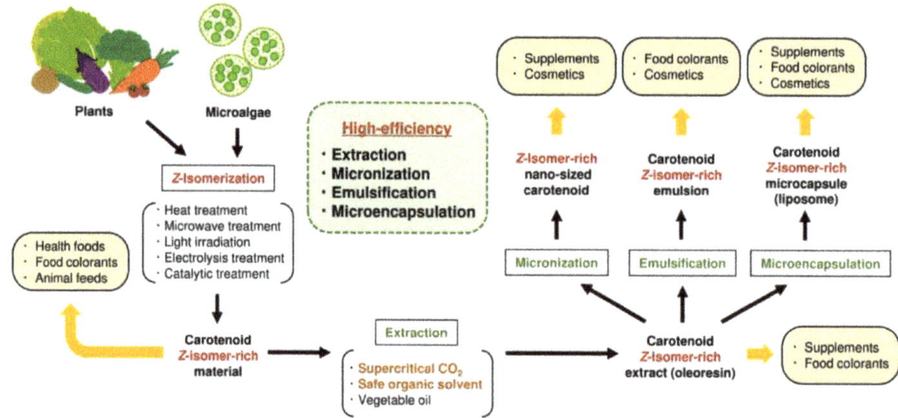

Figure 5. Increased efficiency of carotenoids processing by Z-isomerization and applications of Z-isomer-rich carotenoids materials.

Author Contributions: Conceptualization, M.H., T.F., and M.G.; Writing—Original Draft Preparation, M.H.; Writing—Review & Editing, M.H., H.K., T.H., T.F., and M.G.; Revisions & Final editing, M.H., H.K., T.H., Y.Z., W.D., H.K., R.Y., R.T., T.F., and M.G.

Funding: This work was partly supported by JSPS KAKENHI Grant Number 19K15779 (to M.H.) and the Tatematsu Foundation (to M.H.).

Acknowledgments: The authors are grateful to Tsutomu Kumagai, Chitoshi Kitamura, Yoshinori Inoue, and Munenori Takehara (Department of Materials Science, The University of Shiga Prefecture), and Hiroyuki Ueda, Takuma Higashiura, and Kohei Ichihashi (Innovation Division, Kagome Co., Ltd.) for their kind help and constructive suggestions.

Conflicts of Interest: The authors declare no conflicts of interest.

References

1. Maoka, T. Recent progress in structural studies of carotenoids in animals and plants. *Arch. Biochem. Biophys.* **2009**, *483*, 191–195. [CrossRef] [PubMed]
2. Maoka, T. Carotenoids in marine animals. *Mar. Drugs* **2011**, *9*, 278–293. [CrossRef] [PubMed]
3. Podsędek, A. Natural antioxidants and antioxidant capacity of Brassica vegetables: A review. *LWT-Food Sci. Technol.* **2007**, *40*, 1–11. [CrossRef]
4. Rodriguez-Concepcion, M.; Avalos, J.; Bonet, M.L.; Boronat, A.; Gomez-Gomez, L.; Hornero-Mendez, D.; Limon, M.C.; Meléndez-Martínez, A.J.; Olmedilla-Alonso, B.; Palou, A.; et al. A global perspective on carotenoids: Metabolism, biotechnology, and benefits for nutrition and health. *Prog. Lipid Res.* **2018**, *70*, 62–93. [CrossRef] [PubMed]
5. Yabuzaki, J. Carotenoids database: Structures, chemical fingerprints and distribution among organisms. *Database* **2017**, *2017*, bax004. [CrossRef] [PubMed]
6. Amorim-Carrilho, K.T.; Cepeda, A.; Fente, C.; Regal, P. Review of methods for analysis of carotenoids. *TrAC Trends Anal. Chem.* **2014**, *56*, 49–73. [CrossRef]
7. Krinsky, N.I.; Johnson, E.J. Carotenoid actions and their relation to health and disease. *Mol. Asp. Med.* **2005**, *26*, 459–516. [CrossRef]
8. Ouchi, A.; Aizawa, K.; Iwasaki, Y.; Inakuma, T.; Terao, J.; Nagaoka, S.; Mukai, K. Kinetic study of the quenching reaction of singlet oxygen by carotenoids and food extracts in solution. Development of a singlet oxygen absorption capacity (SOAC) assay method. *J. Agric. Food Chem.* **2010**, *58*, 9967–9978. [CrossRef]
9. Xu, X.R.; Zou, Z.Y.; Huang, Y.M.; Xiao, X.; Ma, L.; Lin, X.M. Serum carotenoids in relation to risk factors for development of atherosclerosis. *Clin. Biochem.* **2012**, *45*, 1357–1361. [CrossRef]
10. Schierle, J.; Bretzel, W.; Bühler, I.; Faccin, N.; Hess, D.; Steiner, K.; Schüep, W. Content and isomeric ratio of lycopene in food and human blood plasma. *Food Chem.* **1997**, *59*, 459–465. [CrossRef]
11. Van Breemen, R.B.; Xu, X.; Viana, M.A.; Chen, L.; Stacewicz-Sapuntzakis, M.; Duncan, C.; Bowen, P.E.; Sharifi, R. Liquid chromatography-mass spectrometry of cis-and all-*trans*-lycopene in human serum and prostate tissue after dietary supplementation with tomato sauce. *J. Agric. Food Chem.* **2002**, *50*, 2214–2219. [CrossRef]
12. Honda, M.; Murakami, K.; Watanabe, Y.; Higashiura, T.; Fukaya, T.; Wahyudiono; Kanda, H.; Goto, M. The *E/Z* isomer ratio of lycopene in foods and effect of heating with edible oils and fats on isomerization of (all-*E*)-lycopene. *Eur. J. Lipid Sci. Technol.* **2017**, *119*, 1600389. [CrossRef]
13. Murakami, K.; Honda, M.; Takemura, R.; Fukaya, T.; Kubota, M.; Wahyudiono; Kanda, H.; Goto, M. The thermal *Z*-isomerization-induced change in solubility and physical properties of (all-*E*)-lycopene. *Biochem. Biophys. Res. Commun.* **2017**, *491*, 317–322. [CrossRef]
14. Honda, M.; Kodama, T.; Kageyama, H.; Hibino, T.; Wahyudiono; Kanda, H.; Goto, M. Enhanced solubility and reduced crystallinity of carotenoids, β-carotene and astaxanthin, by Z-Isomerization. *Eur. J. Lipid Sci. Technol.* **2018**, *120*, 1800191. [CrossRef]
15. Clarke, C.J.; Tu, W.C.; Levers, O.; Bröhl, A.; Hallett, J.P. Green and sustainable solvents in chemical processes. *Chem. Rev.* **2018**, *118*, 747–800. [CrossRef]
16. Procopio, A.; Alcaro, S.; Nardi, M.; Oliverio, M.; Ortuso, F.; Sacchetta, P.; Pieragostino, D.; Sindona, G. Synthesis, biological evaluation, and molecular modeling of oleuropein and its semisynthetic derivatives as cyclooxygenase inhibitors. *J. Agric. Food Chem.* **2009**, *57*, 11161–11167. [CrossRef]
17. Nardi, M.; Sindona, G.; Costanzo, P.; Oliverio, M.; Procopio, A. Eco-friendly stereoselective reduction of α,β-unsaturated carbonyl compounds by Er(OTf)$_3$/NaBH$_4$ in 2-MeTHF. *Tetrahedron* **2015**, *71*, 1132–1135. [CrossRef]
18. Honda, M.; Watanabe, Y.; Murakami, K.; Hoang, N.N.; Wahyudiono; Kanda, H.; Goto, M. Enhanced lycopene extraction from gac (*Momordica cochinchinensis* Spreng.) by the Z-isomerization induced with microwave irradiation pre-treatment. *Eur. J. Lipid Sci. Technol.* **2018**, *120*, 1700293. [CrossRef]
19. Honda, M.; Watanabe, Y.; Murakami, K.; Takemura, R.; Fukaya, T.; Wahyudiono; Kanda, H.; Goto, M. Thermal isomerization pre-treatment to improve lycopene extraction from tomato pulp. *LWT-Food Sci. Technol.* **2017**, *86*, 69–75. [CrossRef]

20. Ono, M.; Honda, M.; Wahyudiono; Yasuda, K.; Kanda, H.; Goto, M. Production of β-carotene nanosuspensions using supercritical CO_2 and improvement of its efficiency by Z-isomerization pre-treatment. *J. Supercrit. Fluids* **2018**, *138*, 124–131. [CrossRef]

21. Watanabe, Y.; Honda, M.; Higashiura, T.; Fukaya, T.; Machmudah, S.; Wahyudiono; Kanda, H.; Goto, M. Rapid and selective concentration of lycopene Z-isomers from tomato pulp by supercritical CO_2 with co-solvents. *Solvent Extr. Res. Dev.* **2018**, *25*, 47–57. [CrossRef]

22. Cooperstone, J.L.; Ralston, R.A.; Riedl, K.M.; Haufe, T.C.; Schweiggert, R.M.; King, S.A.; Timmers, C.D.; Francis, D.M.; Lesinski, G.B.; Clinton, S.K.; et al. Enhanced bioavailability of lycopene when consumed as *cis*-isomers from *tangerine* compared to red tomato juice, a randomized, cross-over clinical trial. *Mol. Nutr. Food Res.* **2015**, *59*, 658–669. [CrossRef]

23. Østerlie, M.; Bjerkeng, B.; Liaaen-Jensen, S. Plasma appearance and distribution of astaxanthin *E/Z* and *R/S* isomers in plasma lipoproteins of men after single dose administration of astaxanthin. *J. Nutr. Biochem.* **2000**, *11*, 482–490. [CrossRef]

24. Müller, L.; Goupy, P.; Fröhlich, K.; Dangles, O.; Caris-Veyrat, C.; Böhm, V. Comparative study on antioxidant activity of lycopene (Z)-isomers in different assays. *J. Agric. Food Chem.* **2011**, *59*, 4504–4511. [CrossRef]

25. Yang, C.; Zhang, H.; Liu, R.; Zhu, H.; Zhang, L.; Tsao, R. Bioaccessibility, cellular uptake, and transport of astaxanthin isomers and their antioxidative effects in human intestinal epithelial Caco-2 cells. *J. Agric. Food Chem.* **2017**, *65*, 10223–10232. [CrossRef]

26. Aman, R.; Schieber, A.; Carle, R. Effects of heating and illumination on *trans-cis* isomerization and degradation of β-carotene and lutein in isolated spinach chloroplasts. *J. Agric. Food Chem.* **2005**, *53*, 9512–9518. [CrossRef]

27. Honda, M.; Takahashi, N.; Kuwa, T.; Takehara, M.; Inoue, Y.; Kumagai, T. Spectral characterisation of Z-isomers of lycopene formed during heat treatment and solvent effects on the *E/Z* isomerisation process. *Food Chem.* **2015**, *17*, 323–329. [CrossRef]

28. Yuan, J.P.; Chen, F. Isomerization of *trans*-astaxanthin to *cis*-isomers in organic solvents. *J. Agric. Food Chem.* **1999**, *47*, 3656–3660. [CrossRef]

29. Honda, M.; Sato, H.; Takehara, M.; Inoue, Y.; Kitamura, C.; Takemura, R.; Fukaya, T.; Wahyudiono; Kanda, H.; Goto, M. Microwave-accelerated Z-isomerization of (all-*E*)-lycopene in tomato oleoresin and enhancement of the conversion by vegetable oils containing disulfide compounds. *Eur. J. Lipid Sci. Technol.* **2018**, *120*, 180060. [CrossRef]

30. Kessy, H.N.; Zhang, L.; Zhang, H. Lycopene (Z)-isomers enrichment and separation. *Int. J. Food Sci. Technol.* **2013**, *48*, 2050–2056. [CrossRef]

31. Zhao, L.; Zhao, G.; Chen, F.; Wang, Z.; Wu, J.; Hu, X. Different effects of microwave and ultrasound on the stability of (all-*E*)-astaxanthin. *J. Agric. Food Chem.* **2006**, *54*, 8346–8351. [CrossRef]

32. Chen, B.H.; Chen, Y.Y. Stability of chlorophylls and carotenoids in sweet potato leaves during microwave cooking. *J. Agric. Food Chem.* **1993**, *41*, 1315–1320. [CrossRef]

33. Murakami, K.; Honda, M.; Takemura, R.; Fukaya, T.; Wahyudiono; Kanda, H.; Goto, M. Effect of thermal treatment and light irradiation on the stability of lycopene with high Z-isomers content. *Food Chem.* **2018**, *250*, 253–258. [CrossRef]

34. Kuki, M.; Koyama, Y.; Nagae, H. Triplet-sensitized and thermal isomerization of all-trans, 7-cis, 9-cis, 13-cis and 15-cis isomers of β-carotene: Configurational dependence of the quantum yield of isomerization via the T_1 state. *J. Phys. Chem.* **1991**, *95*, 7171–7180. [CrossRef]

35. Jensen, N.H.; Nielsen, A.B.; Wilbrandt, R. Chlorophyll *a*-sensitized *trans-cis* photoisomerization of all-*trans*-β-carotene. *J. Am. Chem. Soc.* **1982**, *104*, 6117–6119. [CrossRef]

36. Honda, M.; Igami, H.; Kawana, T.; Hayashi, K.; Takehara, M.; Inoue, Y.; Kitamura, C. Photosensitized *E/Z* isomerization of (all-*E*)-lycopene aiming at practical applications. *J. Agric. Food Chem.* **2014**, *62*, 11353–11356. [CrossRef]

37. Wei, C.C.; Gao, G.; Kispert, L.D. Selected *cis/trans* isomers of carotenoids formed by bulk electrolysis and iron(III) chloride oxidation. *J. Chem. Soc. Perkin Trans.* **1997**, *2*, 783–786. [CrossRef]

38. Gao, G.; Wei, C.C.; Jeevarajan, A.S.; Kispert, L.D. Geometrical isomerization of carotenoids mediated by cation radical/dication formation. *J. Phys. Chem.* **1996**, *100*, 5362–5366. [CrossRef]

39. Wyman, G.M. The *cis-trans* isomerization of conjugated compounds. *Chem. Rev.* **1955**, *55*, 625–657. [CrossRef]

40. Zechmeister, L. Cis-*trans* isomerization and stereochemistry of carotenoids and diphenyl-polyenes. *Chem. Rev.* **1944**, *34*, 267–344. [CrossRef]

41. Molnár, P. Research of the (*E/Z*)-isomerization of carotenoids in Pécs since the 1970s. *Arch. Biochem. Biophys.* **2009**, *483*, 156–164. [CrossRef]

42. Honda, M.; Kageyama, H.; Hibino, T.; Takemura, R.; Goto, M.; Fukaya, T. Enhanced *Z*-isomerization of tomato lycopene through the optimal combination of food ingredients. *Sci. Rep.* **2019**, *9*, 7979. [CrossRef]

43. Yu, J.; Gleize, B.; Zhang, L.; Caris-Veyrat, C.; Renard, C.M.G.C. Heating tomato puree in the presence of lipids and onion: The impact of onion on lycopene isomerization. *Food Chem.* **2019**, *296*, 9–16. [CrossRef]

44. Honda, M.; Kawana, T.; Takehara, M.; Inoue, Y. Enhanced *E/Z* isomerization of (all-*E*)-lycopene by employing iron(III) chloride as a catalyst. *J. Food Sci.* **2015**, *80*, C1453–C1459. [CrossRef]

45. Rajendran, V.; Chen, B.H. Isomerization of β-carotene by titanium tetrachloride catalyst. *J. Chem. Sci.* **2007**, *119*, 253–258. [CrossRef]

46. Yang, C.; Zhang, L.; Zhang, H.; Sun, Q.; Liu, R.; Li, J.; Wu, L.; Tsao, R. Rapid and efficient conversion of all-*E*-astaxanthin to 9*Z*-and 13*Z*-isomers and assessment of their stability and antioxidant activities. *J. Agric. Food Chem.* **2017**, *65*, 818–826. [CrossRef]

47. Hechelski, M.; Ghinet, A.; Louvel, B.; Dufrénoy, P.; Rigo, B.; Daïch, A.; Waterlot, C. From conventional Lewis acids to heterogeneous montmorillonite K10: Eco-friendly plant-based catalysts used as green Lewis acids. *ChemSusChem* **2018**, *11*, 1249–1277. [CrossRef]

48. Nardi, M.; Bonacci, S.; Cariati, L.; Costanzo, P.; Oliverio, M.; Sindona, G.; Procopio, A. Synthesis and antioxidant evaluation of lipophilic oleuropein aglycone derivatives. *Food Funct.* **2017**, *8*, 4684–4692. [CrossRef]

49. Oliverio, M.; Nardi, M.; Costanzo, P.; Di Gioia, M.; Procopio, A. Erbium salts as non-toxic catalysts compatible with alternative reaction media. *Sustainability* **2018**, *10*, 721. [CrossRef]

50. Unlu, N.Z.; Bohn, T.; Francis, D.M.; Nagaraja, H.N.; Clinton, S.K.; Schwartz, S.J. Lycopene from heat-induced *cis*-isomer-rich tomato sauce is more bioavailable than from all-*trans*-rich tomato sauce in human subjects. *Br. J. Nutr.* **2007**, *98*, 140–146. [CrossRef]

51. Boileau, A.C.; Merchen, N.R.; Wasson, K.; Atkinson, C.A.; Erdman, J.W., Jr. *Cis*-lycopene is more bioavailable than *trans*-lycopene in vitro and in vivo in lymph-cannulated ferrets. *J. Nutr.* **1999**, *129*, 1176–1181. [CrossRef]

52. Failla, M.L.; Chitchumroonchokchai, C.; Ishida, B.K. In vitro micellarization and intestinal cell uptake of *cis* isomers of lycopene exceed those of all-*trans* lycopene. *J. Nutr.* **2008**, *138*, 482–486. [CrossRef]

53. Sun, Q.; Yang, C.; Li, J.; Raza, H.; Zhang, L. Lycopene: Heterogeneous catalytic *E/Z* isomerization and in vitro bioaccessibility assessment using a diffusion model. *J. Food Sci.* **2016**, *81*, C2381–C2389. [CrossRef]

54. Gaziano, J.M.; Johnson, E.J.; Russell, R.M.; Manson, J.E.; Stampfer, M.J.; Ridker, P.M.; Frei, B.; Hennekens, C.H.; Krinsky, N.I. Discrimination in absorption or transport of β-carotene isomers after oral supplementation with either all-*trans*- or 9-*cis*-β-carotene. *Am. J. Clin. Nutr.* **1995**, *61*, 1248–1252. [CrossRef]

55. Johnson, E.J.; Krinsky, N.I.; Russell, R.M. Serum response of all-*trans* and 9-*cis* isomers of β-carotene in humans. *J. Am. Coll. Nutr.* **1996**, *15*, 620–624. [CrossRef]

56. Stahl, W.; Schwarz, W.; von Laar, J.; Sies, H. All-*trans* β-carotene preferentially accumulates in human chylomicrons and very low density lipoproteins compared with the 9-*cis* geometrical isomer. *J. Nutr.* **1995**, *125*, 2128–2133. [CrossRef]

57. Stahl, W.; Schwarz, W.; Sies, H. Human serum concentrations of all-trans β- and α-carotene but not 9-*cis* β-carotene increase upon ingestion of a natural isomer mixture obtained from *Dunaliella salina* (Betatene). *J. Nutr.* **1993**, *123*, 847–851. [CrossRef]

58. Tamai, H.; Morinobu, T.; Murata, T.; Manago, M.; Mino, M. 9-*cis* β-Carotene in human plasma and blood cells after ingestion of β-carotene. *Lipids* **1995**, *30*, 493–498. [CrossRef]

59. Erdman Jr, J.W.; Thatcher, A.J.; Hofmann, N.E.; Lederman, J.D.; Block, S.S.; Lee, C.M.; Mokady, S. All-*trans* β-carotene is absorbed preferentially to 9-*cis* β-carotene, but the latter accumulates in the tissues of domestic ferrets (*Mustela putorius puro*). *J. Nutr.* **1998**, *128*, 2009–2013. [CrossRef]

60. Deming, D.M.; Teixeira, S.R.; Erdman Jr, J.W. All-*trans* β-carotene appears to be more bioavailable than 9-*cis* or 13-*cis* β-carotene in gerbils given single oral doses of each isomer. *J. Nutr.* **2002**, *132*, 2700–2708. [CrossRef]

61. During, A.; Hussain, M.M.; Morel, D.W.; Harrison, E.H. Carotenoid uptake and secretion by CaCo-2 cells: β-carotene isomer selectivity and carotenoid interactions. *J. Lipid Res.* **2002**, *43*, 1086–1095. [CrossRef] [PubMed]

62. Yang, C.; Fischer, M.; Kirby, C.; Liu, R.; Zhu, H.; Zhang, H.; Chen, Y.; Sun, Y.; Zhang, L.; Tsao, R. Bioaccessibility, cellular uptake and transport of luteins and assessment of their antioxidant activities. *Food Chem.* **2018**, *249*, 66–76. [CrossRef]

63. Kopec, R.E.; Failla, M.L. Recent advances in the bioaccessibility and bioavailability of carotenoids and effects of other dietary lipophiles. *J. Food Compos. Anal.* **2018**, *68*, 16–30. [CrossRef]

64. Böhm, V.; Puspitasari-Nienaber, N.L.; Ferruzzi, M.G.; Schwartz, S.J. Trolox equivalent antioxidant capacity of different geometrical isomers of α-carotene, β-carotene, lycopene, and zeaxanthin. *J. Agric. Food Chem.* **2002**, *50*, 221–226. [CrossRef] [PubMed]

65. Levin, G.; Yeshurun, M.; Mokady, S. In vivo antiperoxidative effect of 9-*cis* β-carotene compared with that of the all-*trans* isomer. *Nutr. Cancer* **1997**, *27*, 293–297. [CrossRef] [PubMed]

66. Mueller, L.; Boehm, V. Antioxidant activity of β-carotene compounds in different in vitro assays. *Molecules* **2011**, *16*, 1055–1069. [CrossRef] [PubMed]

67. Liu, X.; Osawa, T. *Cis* astaxanthin and especially 9-*cis* astaxanthin exhibits a higher antioxidant activity in vitro compared to the all-*trans* isomer. *Biochem. Biophys. Res. Commun.* **2007**, *357*, 187–193. [CrossRef] [PubMed]

68. Venugopalan, V.; Tripathi, S.K.; Nahar, P.; Saradhi, P.P.; Das, R.H.; Gautam, H.K. Characterization of canthaxanthin isomers isolated from a new soil *Dietzia* sp. and their antioxidant activities. *J. Microbiol. Biotechnol.* **2013**, *23*, 237–245. [CrossRef] [PubMed]

69. Kawee-ai, A.; Kuntiya, A.; Kim, S.M. Anticholinesterase and antioxidant activities of fucoxanthin purified from the microalga *Phaeodactylum tricornutum*. *Nat. Prod. Commun.* **2013**, *8*, 1381–1386. [CrossRef]

70. Zhang, Y.; Fang, H.; Xie, Q.; Sun, J.; Liu, R.; Hong, Z.; Yi, R.; Wu, H. Comparative evaluation of the radical-scavenging activities of fucoxanthin and its stereoisomers. *Molecules* **2014**, *19*, 2100–2113. [CrossRef]

71. Harari, A.; Harats, D.; Marko, D.; Cohen, H.; Barshack, I.; Kamari, Y.; Gonen, A.; Gerber, Y.; Ben-Amotz, A.; Shaish, A. A 9-*cis* β-carotene–enriched diet inhibits atherogenesis and fatty liver formation in LDL receptor knockout mice. *J. Nutr.* **2008**, *138*, 1923–1930. [CrossRef] [PubMed]

72. Harari, A.; Abecassis, R.; Relevi, N.; Levi, Z.; Ben-Amotz, A.; Kamari, Y.; Harats, A.; Shaish, A. Prevention of atherosclerosis progression by 9-cis-β-carotene rich alga *Dunaliella* in apoE-deficient mice. *Biomed. Res. Int.* **2013**, *2013*, 169517. [CrossRef] [PubMed]

73. Relevy, N.Z.; Rühl, R.; Harari, A.; Grosskopf, I.; Barshack, I.; Ben-Amotz, A.; Nir, U.; Gottlieb, H.; Kamari, Y.; Harats, D.; et al. 9-*cis* β-Carotene inhibits atherosclerosis development in female LDLR-/- mice. *Funct. Foods Health Dis.* **2015**, *5*, 67–79.

74. Venugopalan, V.; Verma, N.; Gautam, H.K.; Saradhi, P.P.; Das, R.H. 9-*cis*-Canthaxanthin exhibits higher pro-apoptotic activity than all-*trans*-canthaxanthin isomer in THP-1 macrophage cells. *Free Radic. Res.* **2009**, *43*, 100–105. [CrossRef] [PubMed]

75. Nakazawa, Y.; Sashima, T.; Hosokawa, M.; Miyashita, K. Comparative evaluation of growth inhibitory effect of stereoisomers of fucoxanthin in human cancer cell lines. *J Funct. Foods* **2009**, *1*, 88–97. [CrossRef]

76. Yang, C.; Hassan, Y.I.; Liu, R.; Zhang, H.; Chen, Y.; Zhang, L.; Tsao, R. Anti-inflammatory effects of different astaxanthin isomers and the roles of lipid transporters in the cellular transport of astaxanthin isomers in Caco-2 cell monolayers. *J. Agric. Food Chem.* **2019**, *67*, 6222–6231. [CrossRef]

77. Nagao, A.; Olson, J.A. Enzymatic formation of 9-*cis*, 13-*cis*, and all-*trans* retinals from isomers of β-carotene. *FASEB J.* **1994**, *8*, 968–973. [CrossRef]

78. Schieber, A.; Carle, R. Occurrence of carotenoid *cis*-isomers in food: Technological, analytical, and nutritional implications. *Trends Food Sci. Technol.* **2005**, *16*, 416–422. [CrossRef]

79. Liu, X.; Chen, X.; Liu, H.; Cao, Y. Antioxidation and anti-aging activities of astaxanthin geometrical isomers and molecular mechanism involved in *Caenorhabditis elegans*. *J. Funct. Foods* **2018**, *44*, 127–136. [CrossRef]

80. Fenni, S.; Astier, J.; Bonnet, L.; Karkeni, E.; Gouranton, E.; Mounien, L.; Couturier, C.; Tourniaire, F.; Böhm, V.; Hammou, H.; et al. (all-*E*)-and (5*Z*)-Lycopene display similar biological effects on adipocytes. *Mol. Nutr. Food Res.* **2019**, *63*, 1800788. [CrossRef]

81. Ferruzzi, M.G.; Lumpkin, J.L.; Schwartz, S.J.; Failla, M. Digestive stability, micellarization, and uptake of β-carotene isomers by Caco-2 human intestinal cells. *J. Agric. Food Chem.* **2006**, *54*, 2780–2785. [CrossRef] [PubMed]

82. Honda, M.; Kudo, T.; Kuwa, T.; Higashiura, T.; Fukaya, T.; Inoue, Y.; Kitamura, C.; Takehara, M. Isolation and spectral characterization of thermally generated multi-*Z*-isomers of lycopene and the theoretically preferred pathway to di-Z-isomers. *Biosci. Biotechnol. Biochem.* **2017**, *81*, 365–371. [CrossRef] [PubMed]

83. Jing, C.; Qun, X.; Rohrer, J. HPLC separation of all-*trans*-β-carotene and its iodine-induced isomers using a C30 column. *Thermo Sci.* **2012**, *187*, 1–5.

84. Gamlieli-Bonshtein, I.; Korin, E.; Cohen, S. Selective separation of *cis-trans* geometrical isomers of β-carotene via CO_2 supercritical fluid extraction. *Biotechnol. Bioeng.* **2002**, *80*, 169–174. [CrossRef] [PubMed]

85. Murakami, K.; Honda, M.; Wahyudiono; Kanda, H.; Goto, M. Thermal isomerization of (all-*E*)-lycopene and separation of the Z-isomers by using a low boiling solvent: Dimethyl ether. *Sep. Sci. Technol.* **2017**, *52*, 2573–2582. [CrossRef]

86. Desmarchelier, C.; Borel, P. Overview of carotenoid bioavailability determinants: From dietary factors to host genetic variations. *Trends Food Sci. Technol.* **2017**, *69*, 270–280. [CrossRef]

87. Boileau, T.W.M.; Boileau, A.C.; Erdman, J.W., Jr. Bioavailability of all-*trans* and *cis*-isomers of lycopene. *Exp. Biol. Med.* **2002**, *227*, 914–919. [CrossRef]

88. Hempel, J.; Schädle, C.N.; Leptihn, S.; Carle, R.; Schweiggert, R.M. Structure related aggregation behavior of carotenoids and carotenoid esters. *J. Photochem. Photobiol. A Chem.* **2016**, *317*, 161–174. [CrossRef]

89. Hempel, J.; Amrehn, E.; Quesada, S.; Esquivel, P.; Jiménez, V.M.; Heller, A.; Carle, R.; Schweiggert, R.M. Lipid-dissolved γ-carotene, β-carotene, and lycopene in globular chromoplasts of peach palm (*Bactris gasipaes* Kunth) fruits. *Planta* **2014**, *240*, 1037–1050. [CrossRef]

90. Ben-Amotz, A.; Lers, A.; Avron, M. Stereoisomers of β-carotene and phytoene in the alga *Dunaliella bardawil*. *Plant Physiol.* **1988**, *86*, 1286–1291. [CrossRef]

91. Takehara, M.; Nishimura, M.; Kuwa, T.; Inoue, Y.; Kitamura, C.; Kumagai, T.; Honda, M. Characterization and thermal isomerization of (all-*E*)-lycopene. *J. Agric. Food Chem.* **2014**, *62*, 264–269. [CrossRef] [PubMed]

92. Chasse, G.A.; Mak, M.L.; Deretey, E.; Farkas, I.; Torday, L.L.; Papp, J.G.; Sarma, D.S.R.; Agarwal, A.; Chakravarthi, S.; Agarwal, S.; et al. An ab initio computational study on selected lycopene isomers. *J. Mol. Struct. THEOCHEM* **2001**, *571*, 27–37. [CrossRef]

93. Takehara, M.; Kuwa, T.; Inoue, Y.; Kitamura, C.; Honda, M. Isolation and characterization of (15Z)-lycopene thermally generated from a natural source. *Biochem. Biophys. Res. Commun.* **2015**, *467*, 58–62. [CrossRef] [PubMed]

94. Guo, W.H.; Tu, C.Y.; Hu, C.H. Cis-trans isomerizations of β-carotene and lycopene: A theoretical study. *J. Phys. Chem. B* **2008**, *112*, 12158–12167. [CrossRef] [PubMed]

95. Richelle, M.; Lambelet, P.; Rytz, A.; Tavazzi, I.; Mermoud, A.F.; Juhel, C.; Borel, P.; Bortlik, K. The proportion of lycopene isomers in human plasma is modulated by lycopene isomer profile in the meal but not by lycopene preparation. *Br. J. Nutr.* **2012**, *107*, 1482–1488. [CrossRef] [PubMed]

96. Mohan, S.; Rao, P.R.; Hemachandran, H.; Pullela, P.K.; Tayubi, I.A.; Subramanian, B.; Gothandam, K.M.; Singh, P.; Ramamoorthy, S. Prospects and progress in the production of valuable carotenoids: Insights from metabolic engineering, synthetic biology, and computational approaches. *J. Biotechnol.* **2018**, *266*, 89–101.

97. Mussagy, C.U.; Winterburn, J.; Santos-Ebinuma, V.C.; Pereira, J.F.B. Production and extraction of carotenoids produced by microorganisms. *Appl. Microbiol. Biotechnol.* **2019**, *103*, 1095–1114. [CrossRef]

98. Rammuni, M.N.; Ariyadasa, T.U.; Nimarshana, P.H.V.; Attalage, R.A. Comparative assessment on the extraction of carotenoids from microalgal sources: Astaxanthin from *H. pluvialis* and β-carotene from *D. salina*. *Food Chem.* **2019**, *277*, 128–134. [CrossRef]

99. Sathasivam, R.; Ki, J.S. A review of the biological activities of microalgal carotenoids and their potential use in healthcare and cosmetic industries. *Mar. Drugs* **2018**, *16*, 26. [CrossRef]

100. Chuyen, H.V.; Nguyen, M.H.; Roach, P.D.; Golding, J.B.; Parks, S.E. Gac fruit (*Momordica cochinchinensis* Spreng.): A rich source of bioactive compounds and its potential health benefits. *Int. J. Food Sci. Technol.* **2015**, *50*, 567–577. [CrossRef]

101. Vuong, L.T.; Franke, A.A.; Custer, L.J.; Murphy, S.P. *Momordica cochinchinensis* Spreng. (gac) fruit carotenoids reevaluated. *J. Food Compos. Anal.* **2006**, *19*, 664–668. [CrossRef]

102. Kha, T.C.; Nguyen, M.H.; Roach, P.D.; Stathopoulos, C.E. Effects of Gac aril microwave processing conditions on oil extraction efficiency, and β-carotene and lycopene contents. *J. Food Eng.* **2013**, *117*, 486–491. [CrossRef]

103. Martins, N.; Ferreira, I.C. Wastes and by-products: Upcoming sources of carotenoids for biotechnological purposes and health-related applications. *Trends Food Sci. Technol.* **2017**, *62*, 33–48. [CrossRef]

104. Saini, R.K.; Keum, Y.S. Carotenoid extraction methods: A review of recent developments. *Food Chem.* **2018**, *240*, 90–103. [CrossRef] [PubMed]

105. Luengo, E.; Álvarez, I.; Raso, J. Improving carotenoid extraction from tomato waste by pulsed electric fields. *Front. Nutr.* **2014**, *1*, 12. [CrossRef] [PubMed]

106. Jaeschke, D.P.; Menegol, T.; Rech, R.; Mercali, G.D.; Marczak, L.D.F. Carotenoid and lipid extraction from *Heterochlorella luteoviridis* using moderate electric field and ethanol. *Proc. Biochem.* **2016**, *51*, 1636–1643. [CrossRef]

107. Altemimi, A.; Lakhssassi, N.; Baharlouei, A.; Watson, D.; Lightfoot, D. Phytochemicals: Extraction, isolation, and identification of bioactive compounds from plant extracts. *Plants* **2017**, *6*, 42. [CrossRef] [PubMed]

108. Chuyen, H.V.; Nguyen, M.H.; Roach, P.D.; Golding, J.B.; Parks, S.E. Microwave-assisted extraction and ultrasound-assisted extraction for recovering carotenoids from Gac peel and their effects on antioxidant capacity of the extracts. *Food Sci. Nutr.* **2018**, *6*, 189–196. [CrossRef] [PubMed]

109. Song, J.; Yang, Q.; Huang, W.; Xiao, Y.; Li, D.; Liu, C. Optimization of trans lutein from pumpkin (*Cucurbita moschata*) peel by ultrasound-assisted extraction. *Food Bioprod. Proc.* **2018**, *107*, 104–112. [CrossRef]

110. Affandi, M.M.M.; Julianto, T.; Majeed, A.B.A. Enhanced oral bioavailability of astaxanthin with droplet size reduction. *Food Sci. Technol. Res.* **2012**, *18*, 549–554. [CrossRef]

111. Vishwanathan, R.; Wilson, T.A.; Nicolosi, R.J. Bioavailability of a nanoemulsion of lutein is greater than a lutein supplement. *Nano Biomed. Eng.* **2009**, *1*, 57–73. [CrossRef]

112. Karam, M.C.; Petit, J.; Zimmer, D.; Djantou, E.B.; Scher, J. Effects of drying and grinding in production of fruit and vegetable powders: A review. *J. Food Eng.* **2016**, *188*, 32–49. [CrossRef]

113. De Paz, E.; Martín, Á.; Estrella, A.; Rodríguez-Rojo, S.; Matias, A.A.; Duarte, C.M.; Cocero, M.J. Formulation of β-carotene by precipitation from pressurized ethyl acetate-on-water emulsions for application as natural colorant. *Food Hydrocoll.* **2012**, *26*, 17–27. [CrossRef]

114. An, Y.; Sun, Y.; Zhang, M.; Adhikari, B.; Li, Z. Effect of ball milling time on physicochemical properties of *Cordyceps militaris* ultrafine particles. *J. Food Proc. Eng.* **2019**, e13065. [CrossRef]

115. Sahena, F.; Zaidul, I.S.M.; Jinap, S.; Karim, A.A.; Abbas, K.A.; Norulaini, N.A.N.; Omar, A.K.M. Application of supercritical CO_2 in lipid extraction—A review. *J. Food Eng.* **2009**, *95*, 240–253. [CrossRef]

116. Zuknik, M.H.; Norulaini, N.N.; Omar, A.M. Supercritical carbon dioxide extraction of lycopene: A review. *J. Food Eng.* **2012**, *112*, 253–262. [CrossRef]

117. Kodama, T.; Honda, M.; Takemura, R.; Fukaya, T.; Uemori, C.; Wahyudiono; Kanda, H.; Goto, M. Effect of the Z-isomer content on nanoparticle production of lycopene using solution-enhanced dispersion by supercritical fluids (SEDS). *J. Supercrit. Fluids* **2018**, *133*, 291–296. [CrossRef]

118. Martín, A.; Cocero, M.J. Micronization processes with supercritical fluids: Fundamentals and mechanisms. *Adv. Drug Del. Rev.* **2008**, *60*, 339–350.

119. Mattea, F.; Martín, Á.; Cocero, M.J. Carotenoid processing with supercritical fluids. *J. Food Eng.* **2009**, *93*, 255–265. [CrossRef]

120. Kodama, T.; Honda, M.; Machmudah, S.; Wahyudiono; Kanda, H.; Goto, M. Crystallization of all *trans*-β-carotene by supercritical carbon dioxide antisolvent via co-axial nozzle. *Eng. J.* **2018**, *22*, 25–38. [CrossRef]

121. Kaga, K.; Honda, M.; Adachi, T.; Honjo, M.; Wahyudiono; Kanda, H.; Goto, M. Nanoparticle formation of PVP/astaxanthin inclusion complex by solution-enhanced dispersion by supercritical fluids (SEDS): Effect of PVP and astaxanthin Z-isomer content. *J. Supercrit. Fluids* **2018**, *136*, 44–51. [CrossRef]

122. Nerome, H.; Machmudah, S.; Wahyudiono; Fukuzato, R.; Higashiura, T.; Youn, Y.S.; Lee, Y.W.; Goto, M. Nanoparticle formation of lycopene/β-cyclodextrin inclusion complex using supercritical antisolvent precipitation. *J. Supercrit. Fluids* **2013**, *83*, 97–103. [CrossRef]

123. Boonnoun, P.; Nerome, H.; Machmudah, S.; Goto, M.; Shotipruk, A. Supercritical anti-solvent micronization of chromatography purified marigold lutein using hexane and ethyl acetate solvent mixture. *J. Supercrit. Fluids* **2013**, *80*, 15–22. [CrossRef]

124. Miguel, F.; Martin, A.; Gamse, T.; Cocero, M.J. Supercritical anti solvent precipitation of lycopene: Effect of the operating parameters. *J. Supercrit. Fluids* **2006**, *36*, 225–235. [CrossRef]

125. Tavares-Cardoso, M.A.; Antunes, S.; van Keulen, F.; Ferreira, B.S.; Geraldes, A.; Cabral, J.; Palavra, A.M. Supercritical antisolvent micronisation of synthetic all-*trans*-β-carotene with tetrahydrofuran as solvent and carbon dioxide as antisolvent. *J. Chem. Technol. Biotechnol.* **2009**, *84*, 215–222. [CrossRef]

126. Leong, H.Y.; Show, P.L.; Lim, M.H.; Ooi, C.W.; Ling, T.C. Natural red pigments from plants and their health benefits: A review. *Food Rev. Int.* **2018**, *34*, 463–482. [CrossRef]
127. Coultate, T.; Blackburn, R.S. Food colorants: Their past, present and future. *Color. Technol.* **2018**, *134*, 165–186. [CrossRef]
128. Anarjan, N.; Tan, C.P. Effects of selected polysorbate and sucrose ester emulsifiers on the physicochemical properties of astaxanthin nanodispersions. *Molecules* **2013**, *18*, 768–777. [CrossRef]
129. Spernath, A.; Aserin, A. Microemulsions as carriers for drugs and nutraceuticals. *Adv. Colloid Interface Sci.* **2006**, *128*, 47–64. [CrossRef]
130. Salvia-Trujillo, L.; Qian, C.; Martín-Belloso, O.; McClements, D.J. Influence of particle size on lipid digestion and β-carotene bioaccessibility in emulsions and nanoemulsions. *Food Chem.* **2013**, *141*, 1472–1480. [CrossRef]
131. Chu, B.S.; Ichikawa, S.; Kanafusa, S.; Nakajima, M. Preparation of protein-stabilized β-carotene nanodispersions by emulsification–evaporation method. *J. Am. Oil Chem. Soc.* **2007**, *84*, 1053–1062. [CrossRef]
132. De Paz, E.; Martín, Á.; Mateos, E.; Cocero, M.J. Production of water-soluble β-carotene micellar formulations by novel emulsion techniques. *Chem. Eng. Process. Process Intensif.* **2013**, *74*, 90–96. [CrossRef]
133. Tan, C.P.; Nakajima, M. β-Carotene nanodispersions: Preparation, characterization and stability evaluation. *Food Chem.* **2005**, *92*, 661–671. [CrossRef]
134. Lakshmi, P.; Kumar, G.A. Nanosuspension technology: A review. *Int. J. Pharm. Sci* **2010**, *2*, 35–40.
135. Ezhilarasi, P.N.; Karthik, P.; Chhanwal, N.; Anandharamakrishnan, C. Nanoencapsulation techniques for food bioactive components: A review. *Food Bioprocess. Technol.* **2013**, *6*, 628–647. [CrossRef]
136. Silva, H.D.; Cerqueira, M.A.; Souza, B.W.S.; Ribeiro, C.; Avides, M.C.; Quintas, M.A.C.; Coimbra, J.S.R.; Carneiro-da-Cunha, M.G.; Vicente, A.A. Nanoemulsions of β-carotene using a high-energy emulsification–evaporation technique. *J. Food Eng.* **2011**, *102*, 130–135. [CrossRef]

MDPI

St. Alban-Anlage 66

4052 Basel

Switzerland

Tel. +41 61 683 77 34

Fax +41 61 302 89 18

www.mdpi.com

Molecules Editorial Office

E-mail: molecules@mdpi.com

www.mdpi.com/journal/molecules